Advances in Earthquake Research

Advances in Earthquake Research

Hammad Hussain Awan

ARCLER
PRESS

www.arclerpress.com

Advances in Earthquake Research

Hammad Hussain Awan

Arcler Press

2010 Winston Park Drive,

2nd Floor

Oakville, ON L6H 5R7

Canada

www.arclerpress.com

Tel: 001-289-291-7705
001-905-616-2116
Fax: 001-289-291-7601
Email: orders@arclereducation.com

ISBN: 978-1-77361-227-0 (Hardcover)

ABOUT THE AUTHOR

Hammad Hussain Awan

Hammad Hussain Awan (Ph.D Student) specializes in research in field of civil engineering which encompasses various subjects including Earthquake Research and Prediction. Key areas of interest are studying various phenomena involved in stability of different structures due to various natural external forces such as earthquakes, wind, snow etc.

Secondary interests also include research in evaluation of traffic safety for road infrastructure designs using different techniques such as Driving Simulator.

TABLE OF CONTENTS

PREFACE

Advances in Earthquake Research is a combination of multidisciplinary learning in a few regions of essential sciences and science-based designing with a definitive objective of diminishing the seismic dangers to financially worthy levels. A primary extensive book covering different parts of Earthquake Engineering has been under discussion since 1900. There have been late advances and new advancements in Earthquake Engineering on an extensive variety of themes, from geosciences and geotechnical designing to present day execution based Earthquake Engineering. These advances are distributed in logical and specialized diaries and reports or displayed at national and global meetings. This book has been composed with the goal of exhibiting advances in logical information on different Earthquake Engineering subjects in a solitary volume. In spite of the fact that it has not been composed in a conventional course reading design, this book can serve as a guide for educators, graduate understudies and field engineers. We trust it will add to the instructing of current Earthquake Advances and its applications to practical cases, and in addition to the definition and development of research projects.

The seven parts in this book can be assembled into the accompanying advances in Seismic research:

- Historical advancement of Earthquake Research and structure of earth (Chapter 1)
- Scientific study of earthquakes and the propagation of elastic waves through the Earth. (Chapter 2)
- Engineering portrayals of ground movement, and geotechnical dangers (Chapter 2)
- Topography gives the quantitative establishment on which different parts of seismic peril alleviation are based. (Chapter 3)
- Solid ground shaking underneath the establishment of a structure is the essential impact of an Earthquake. (Chapter 4)
- Response of Simple Structures to Earthquake w.r.t understanding and controlling seismic tremor reaction of structures are Stiffness, Strength and Ductility. (Chapter 5)

- Examination of structures for Dynamic Response Analysis. (Chapter 6)
- Innovative systems and methods in Earthquake Research (Chapter 7)

The multidisciplinary variety of earthquake research makes it extremely hard to cover the points of interest of all the logical and building perspectives required in current Earthquake Research in a solitary volume. Due to space requirements and furthermore to the presence of other elegantly composed books and handbooks, I have not secured some critical Earthquake Research points in this book. These incorporate an expound exchange of direct auxiliary progression; design contemplations; seismic conduct what's more, outline of helps and modern offices; hazard administration; and social, efficient what's more, political arranging. The expansiveness of Earthquake Research makes it unthinkable for one individual to legitimately expound on all pertinent themes.

1
CHAPTER

NATURE OF EARTHQUAKES

INTRODUCTION

An earthquake is the shaking of the earth's surface. There can be various reasons that can result in an earthquake, varying from sudden release of energy from earth's interior to volcanic eruptions and many more. An earthquake of high magnitude is a frightening experience. It can cause enormous damages to human life, homes and goods in a matter of only a few minutes. Compared to any other natural disaster, the earthquake is the most dangerous, because it weakens the basic stability of human existence and the confidence that this stability is under control. When this stability is destroyed by the violence of an earthquake, fright and panic can become unbearable.

HISTORY OF EARTHQUAKES

Despite the fact that seismic tremors have been occurring since the earth was formed; the logical investigation of quakes is relatively new. Until the start of eighteenth century, only a couple of accurate portrayals of seismic tremors were recorded, and the normal reason for quakes was minimally understood. Numerous hypotheses were created by the individuals who looked for normal causes that appear to be invented in current time. One well known hypothesis was that seismic tremors were brought on because of ejection to let some circulation into of openings somewhere down inside of Earth.

The documented seismic tremor that can be found in the early history of humankind with definite data occurred in China in 1177 B.C. The Chinese tremor index portrays that a few dozen expansive seismic tremors occurred in China amid the following couple of thousand years. Quakes in Europe can be found in history as early as 580 B.C., yet the soonest for which generally point by point data can be discovered occurred in the mid-sixteenth century. The most punctual known seismic tremors in the Americas were in Mexico in the late fourteenth century and in Peru in 1471, however portrayals of the impacts were not all around archived. By the seventeenth century, depictions of the impacts of tremors were being distributed far and wide, in spite of the fact that these records were frequently overstated or mutilated. The most broadly felt seismic tremors in the North America's written history were an arrangement that occurred in 1811-1812 close New Madrid, Missouri. An incredible seismic tremor, whose size is assessed to be around 8, occurred on the morning of December 16, 1811. Another awesome tremor occurred on January 23, 1812, and a third, the most grounded yet, on February 7, 1812. Post-quake tremors were about constant between these extraordinary seismic tremors and proceeded for quite a long time subsequently. These quakes were felt by individuals as far away as Boston and Denver. Since the most exceptional impacts were not in thickly populated locale, the decimation to human life and property was pretty much nothing. Happening of only one of these colossal seismic tremors in a similar range today would decimate and harm a large number of individuals, structures and different structures worth billions of dollars.

The San Francisco earthquake of 1906, shown in Fig.1.1, was one of the most destructive in the recorded history of North America, the earthquake and the fire that followed killed nearly 700 people and left the city in ruins. The Alaska earthquake of March 27, 1964, was of greater magnitude than the San Francisco earthquake; it released perhaps twice as much energy and was felt over an area of almost 500,000 square miles.

The ground motion near the epi-centre was so violent that the tops of some trees were snapped off. One hundred and fourteen people (some as far away as California) died as a result of this earthquake, but loss of life and property in Alaska would have been far greater had Alaska been more densely populated.

Figure 1.1: A Dramatic Picture of Destruction of the great 1906 San Francisco earthquake

DYNAMIC EARTH STRUCTURE

The inner structure of the Earth assumes an essential part keeping in mind the end goal to comprehend the major seismic movement around the globe. The creation of Earth's inside might be considered to comprise of three concentric layers (Fig. 1.2). The deepest layer of the Earth is called center and it is fundamentally made out of iron. The center can further be partitioned into two separate parts: inward and external center. The internal center is in strong state though the external center is in fluid state. The second layer is called mantle and is between the outside layer (peripheral layer of the earth) and the center. The unexpected changes in the spread speed of seismic waves (Fig. 1.3) separate the mantle, the external center and the inward center. The sudden variety in the seismic wave speed near the crustal surface is because of "Moho irregularity" (perceived by the Croatian seismologist Mohorovicˇic' in 1909) and it is acknowledged as the limit between the mantle and the hull (Fig. 1.2).

The internal structure of the Earth plays a fundamental part in remembering the true objective to research the major seismic development around the world. The formation of Earth's inside may be considered to contain three concentric layers (Fig. 1.2). The most profound layer of the Earth is called the focus and it is, in a general sense, made out of iron.

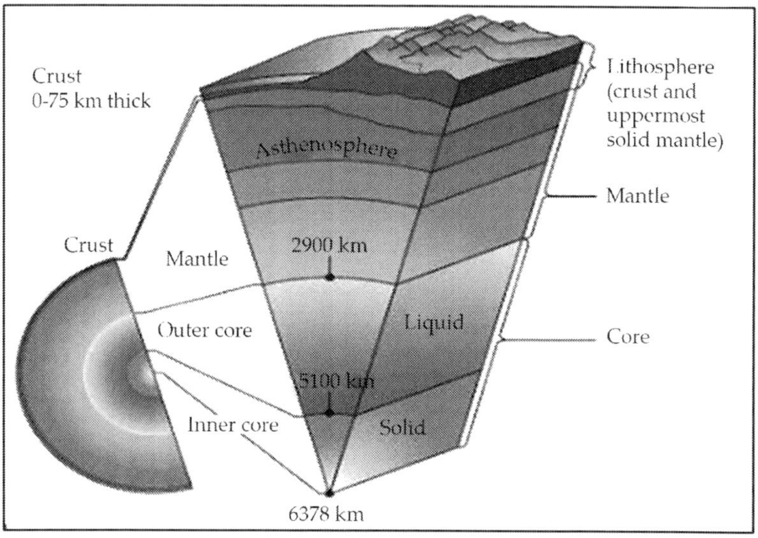

Figure 1.2: Crust of Earth-Inner Side

The inside can be further divided into two separate parts: inner and outer focus. The interior focus is in solid state however the outside focus is in liquid state. The second layer is called mantle and is between the outside layer (fringe layer of the earth) and the middle. The unforeseen changes in the spread speed of seismic waves (Fig. 1.3) isolate the mantle, the outside focus and the internal focus. The sudden assortment in the seismic wave speed close to the crustal surface is a direct result of "Moho abnormality" (seen by the Croatian seismologist Mohorovicˇic' in 1909) and it is recognized as the farthest point between the mantle and the frame (Fig. 1.2). Fig.1.2: Internal Core-Structure of Earth

The outside layer thickness shifts around 7 km under the seas to normal 30 km under the landmasses. The covering gets much thicker under the mountain ranges. The covering has basaltic structure under the seas while it is primarily involved basalt and rock under the mainland. The lithosphere and asthenosphere are the two peripheral layers of the Earth that are characterized as far as material quality and solidness (Fig. 1.2).

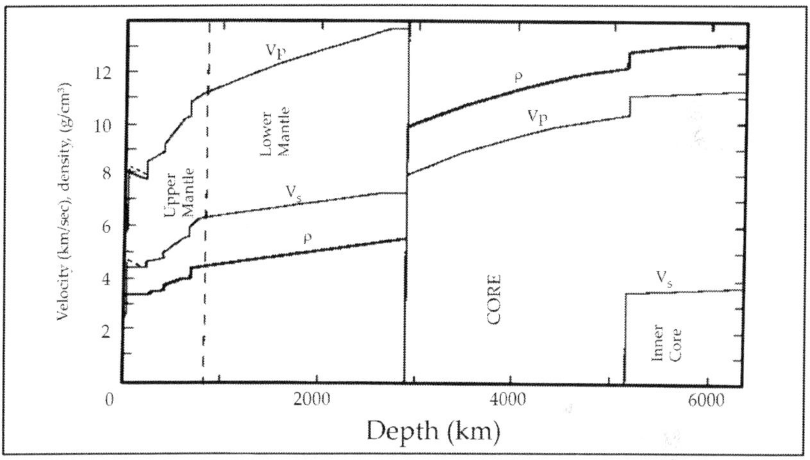

Figure 1.3: The Preliminary Reference Earth Model (PREM) of compressional velocity (Vp), shear velocity (Vs), and density (ρ) as a function of depth in the Earth.

The lithosphere is unbending and moderately more grounded among these two. It is predominantly shaped of the covering and the peripheral piece of the mantle. The thickness of lithosphere is around 125 km, underneath the lithosphere lies the asthenosphere and it shapes essentially the frail piece of the mantle (a gentler layer) that can disfigure through crawl. The condition of asthenosphere at a few spots can be depicted as plastic. Consequently, the lithosphere can be considered to glide over the asthenosphere in such places. The inside of the Earth is in steady movement brought on by warmth. The warmth is created by the radioactivity inside the center. The temperature slope over the Earth sets up a warmth stream towards the surface from the external center. The warmth is exchanged by the component of convection. Convection streams inside the asthenosphere are the reason for developments in the lithospheric plates which are otherwise called structural plates (Fig. 1.4).

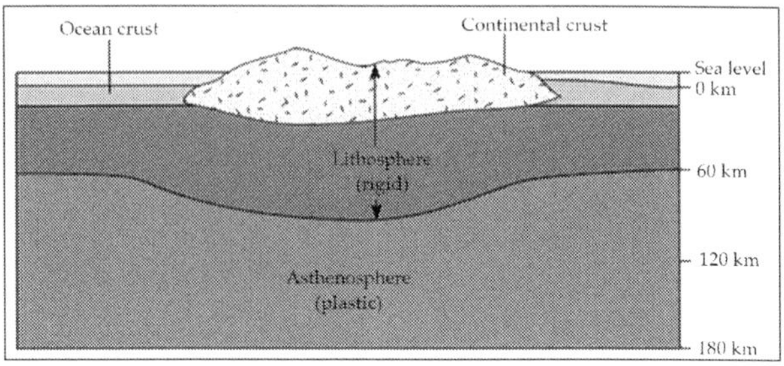

Figure 1.4: Internal Surface Structure of Earth

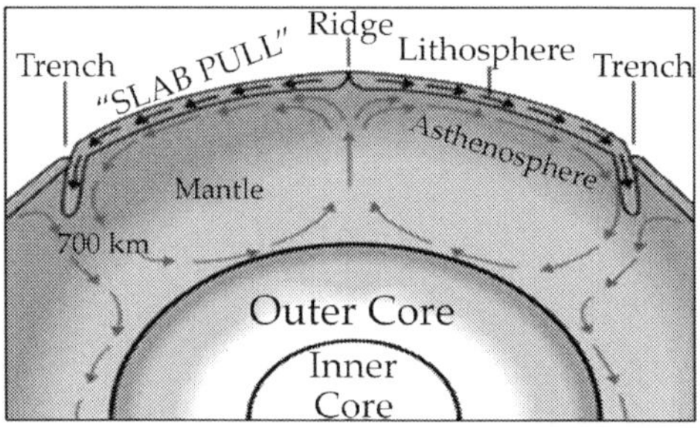

Figure 1.5: Mantle convection cells: Earth's heat from the interior to the surface

The movement of these plates results in two slabs diverging from each other, or converging to each other shown in Fig.1.5. When two slabs converge to each other, they collide and one slab descends beneath the other one. These converging and diverging movements of lithosphere plates result in formation of trenches and ridges. Trenches are deep part of the oceanic crust and are formed when one plate is sub-ducted under the other. Ridges are formed due to diverging movements of lithospheric plates. Ridges are also formed deep in the ocean due to rising of molten magma through the opening formed by the divergence of plates and result in volcanic eruptions(Sucuoğlu & Akkar, 2014).

Continental Drift

The physical process of movements in lithospheric plates due to convection currents in mantle also describes the continuous motion of the continents. According to geologists, 225 million years ago all of the continents consisted of a single landmass which they name as Pangaea. Due to the movements in interior layers of earth this continent broke up initially forming two continents, Laurasia and Gondwanaland, about 200 million years ago. By 135 million years ago, Laurasia had split into the continents of North America and Eurasia, and Gondwanaland had divided into the continents of India, South America, Africa, Antarctica and Australia. These continents have continued to move and have come to their current configuration, including the collision of India with Eurasia about 50 million years ago. The entire process is illustrated in Fig. 1.6. Geologists started looking for explanations about the continental movements in second half of the 20th century. Richard Field, one of amongst these, studied the geology of the oceanic floor. The discovery of mountain ridges along the major oceans and observing dense seismic activity along these oceanic ridges revealed that they are under continuous deformation. This is shown in Fig. 1.6. In 1960, a theory was proposed by Harry Hess that ocean floor is being formed continuously by the rising of magma from the mantle through the openings of the oceanic ridges (Fig. 1.8).

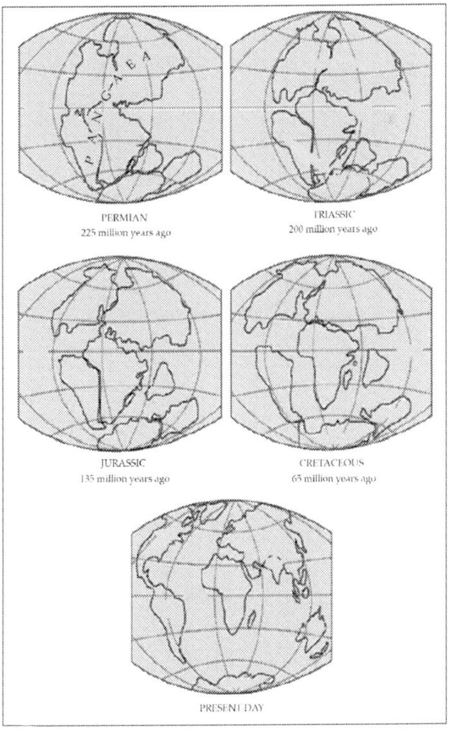

Figure 1.6: Continental Drift Myth- Past 225M Year

(Ref: http://pubs.usgs.gov/gip/dynamic/historical.html)

The magma spreading out from these openings pushes apart two sides of the ridge. This phenomenon separates the two tectonic plates from each other as in the case of African and South American continents. Today, the continuous formation of oceanic floor is still diverging these two continents from each other. A German meteorologist Alfred Wegener first documented the separation of African and South American continents in 1915. He compared the geological structures, mineral deposits and fossils of both flora and fauna from the two sides of the Atlantic Ocean. His hypothesis on continental drift was not accepted by the scientific community during that time as he failed to explain the physical process behind the separation of the continents.

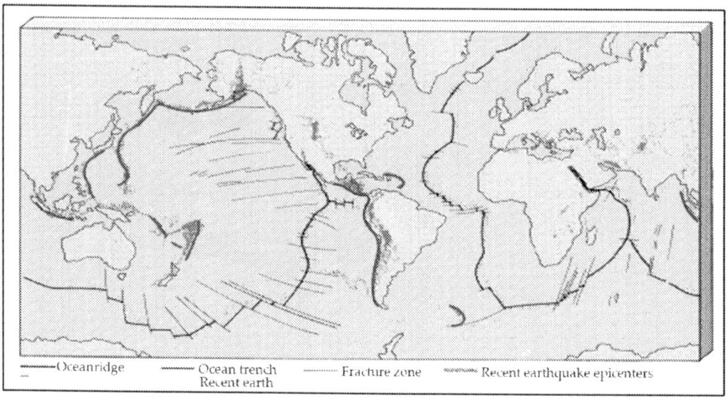

Figure 1.7: Worldwide System of Transverse Fracture Zones.

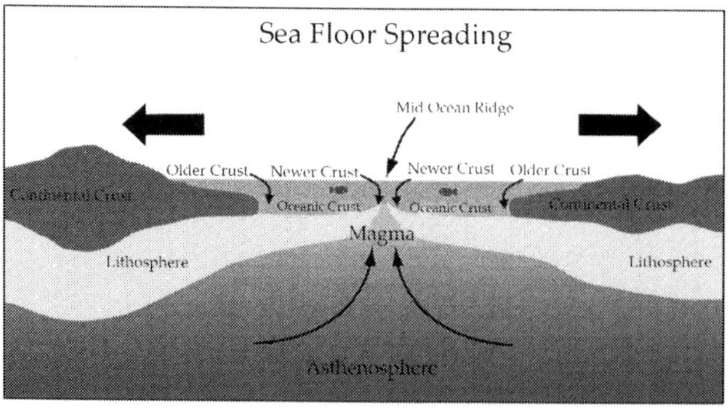

Figure 1.8: Concept of Sea Floor Spreading

An occurrence of another phenomenon which consumes the older crust at the same rate at that new crust is being formed continuously at the mid-oceanic ridges keeps the earth's crust in equilibrium and prevents its expansion.

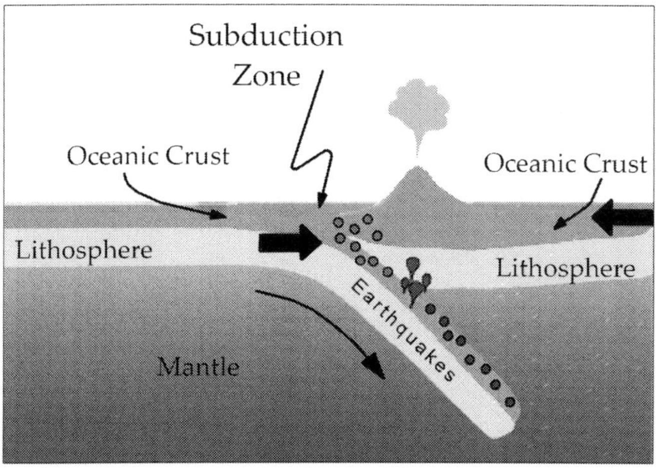

Figure 1.9: Subduction Zone Mechanism

This phenomenon is called subduction and it occurs where two plates collide and one is pushed down below the other (Fig. 1.8). Along with intense seismic activity, volcanic activity is another characteristic associated with subduction regions (Sucuoğlu & Akkar, 2014).

Theory of Global Plate Tectonics

The evidence of mid-oceanic ridges and subduction regions along with high seismic activity at those regions led to the development of global plate tectonics theory (e.g., Isacks et al. 1968; McKenzie 1968). According to this theory, the Earth's lithosphere is divided into a number of lithospheric slabs which are called tectonic plates. These tectonic plates move relative to each other due to convection currents formed in the mantle. The manner in which tectonic plates interact with each other at their boundaries is illustrated in Fig. 1.10. As a result of this interaction, boundaries are formed which can be classified as converging, diverging and transform boundary. At the ocean ridges where plates move apart from each are called divergent plate boundaries.

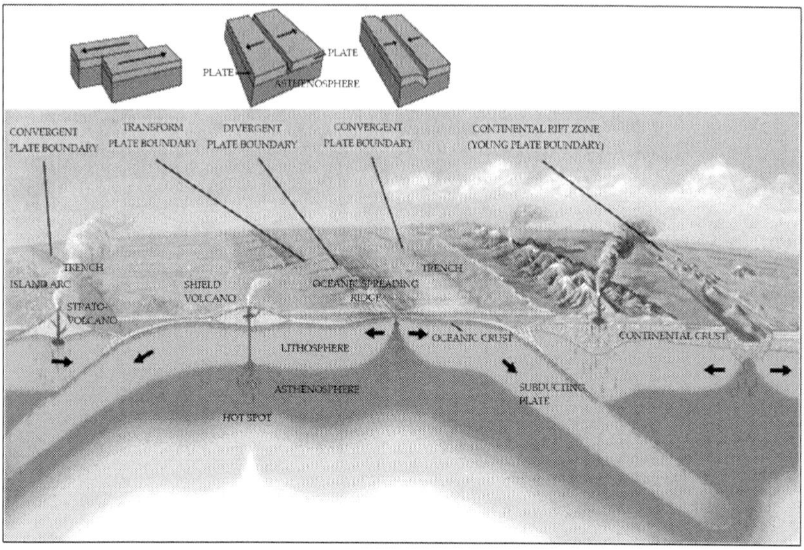

Figure 1.10: Plate Movement during Sea Floor Spreading

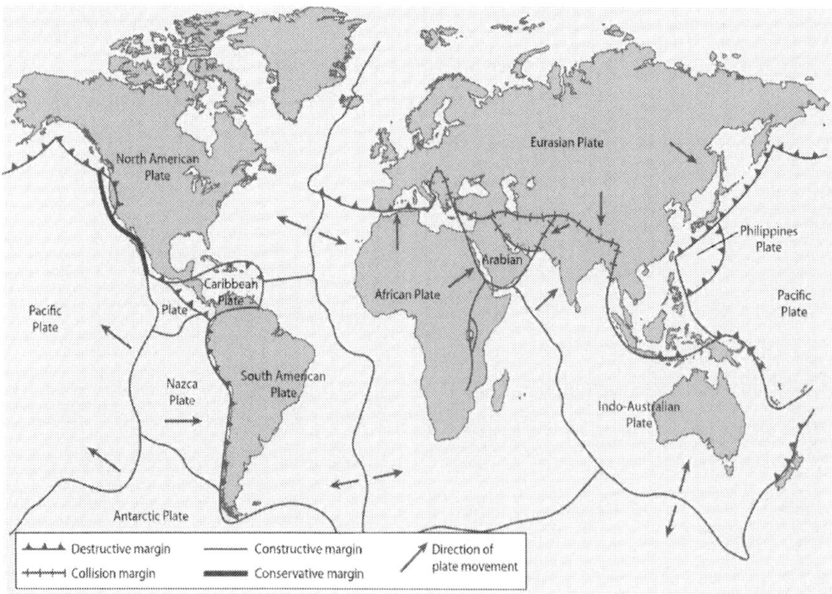

Figure 1.11: Fault Lines on Plate boundaries

At convergent boundaries two plates collide. As a result, one plate is usually driven below the other through the process of subduction. One of the examples of plate subduction is the subduction of oceanic plate below continental plate along the Pacific coast of South and Central America. Due to the process of sea-floor spreading, oceanic crust is denser than the continental crust. Hence, when oceanic and continental crust collide, oceanic crust is sub-ducted below due to its high density.

In case of collision of two continental plates, lithosphere along the plate boundary gets deformed and its thickness is enormously increased (e.g., the Himalayas). In case of horizontal movement of two plates transform (or trans-current) boundaries are created. Such boundaries can be seen along long and well-defined faults such as the San Andreas Fault in California, which is the boundary between the North American and Pacific plates. Figure 1.11 shows the distribution of three major plate boundaries around the globe.

Explanations of most of the earthquakes can be justified by the relative motion of tectonic plates. Almost all earthquakes around the world are located along the tectonic plate boundaries. They are called as inter-plate earthquakes and are represented in Fig.1.12. The circumference of Pacific Ocean where generally subduction process occurs lies between the oceanic and continental slabs and is the most active boundary region. The Mediterranean Sea and surroundings including the Azores islands in the Atlantic Ocean and a significant portion of Asia constitute the other plate boundary regions generating inter-plate earthquakes. Reasons for these inter-plate earthquakes in regions mentioned above are all types of tectonic plate interactions i.e. convergent, divergent and transform.

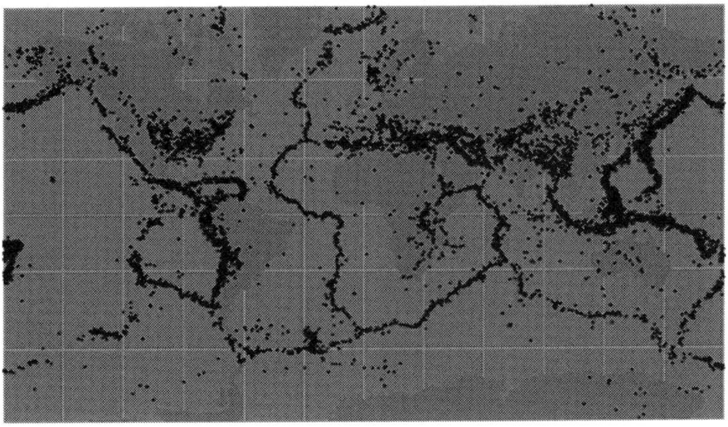

Figure 1.12: Transformation of Fault lines

The earthquakes that occur away from plate boundaries are called as intraplate earthquakes. Most of these earthquakes have occurred in northeast America, Australia, central India and northeast Brazil. Factors causing intraplate earthquakes are different than interplate earthquakes. Interplate earthquakes are caused due to high deformations along the plate boundaries. This is not the case in intraplate earthquakes as clear boundaries do not exist between plates. Areas where intraplate earthquakes are caused are known as stable continental regions. The frequency of occurrence of intraplate earthquakes is less than that of interplate earthquakes but their magnitudes can be very high. Three such earthquakes occurred during the period of December 1811 and February 1812 in the New Madrid Zone in central United States. Their magnitude was between 7.5 and 7.7. The New Madrid Zone is one of the famous stable continental regions in the world and the three earthquakes mentioned previously are among the top largest earthquakes in North America during the past 200 years.

EARTHQUAKE PROCESS AND FAULTS

On-going processes in the earth's interior, discussed earlier, explain the relative movements of tectonic plates and factors causing them. These relative movements result in occurrence of earthquakes along the boundaries of these tectonic plates. After the San Francisco earthquake in 1906, Reid introduced a theory which explained actual mechanism of earthquakes. This theory is known as "elastic rebound theory" and is the first physically justified explanation on the relation of earthquakes with formation of geological faults. According to this theory, elastic stresses are stored within earth's interior. With passage of time, these stresses gradually keep increasing in magnitude and when released suddenly, result in earthquake. As a consequence of this sudden release of energy, faults are formed. Earthquake of San Francisco (1906) was studied in detail. A lateral horizontal slip of the fault was measured according to the fences or roads. It was found that the slip between two sides of the faults varied between 2 to 4m, comparing the measurements taken after the earthquake to those taken before it showed that the two sides of the fault had been in continuous motion even before the occurrence of the earthquake. Even the direction of motion of slip before and after the earthquake was found to be same. Harry Fielding Reid proposed the elastic bound theory based on these observations. Fig.1.17 explains the mechanism of earthquake occurrence according to this theory. As stress develops on opposite side of a fault, the energy starts to accumulate between them. Gradually, when these stresses exceed internal strength of the sides, an unexpected movement across the fault is developed releasing the stored energy resulting in formation of the fault. Once the energy is released, the opposite sides of the fault return to their un-deformed shape.

The elastic rebound theory is the first theory describing fault rupture as the cause of strong earthquake. Before this, the earthquakes were believed to be the cause of formation of fault ruptures. All earthquakes are caused due to the fracture between geological faults except those caused by the unexpected and massive movements of magma. These exceptions are known as volcanic earthquakes. Once a fracture has been developed at a certain point, it starts to propagate along the fault plane very rapidly.

Studying fault ruptures can be very complex sometimes but their overall behaviour along with their different characteristics can be understood easily by idealizing two rectangular blocks as two opposite sides of the fault plane. This is illustrated in Fig. 1.17. Upper part of the fault is termed as hanging wall while the lower part as footwall. The hanging wall moves with respect to footwall. The angle between the fault plane and horizontal ground surface is termed as dip angle δ. It is measured downwards from the horizontal surface and its values lie between 0 and 90°. The strike Φ is the angle measured clockwise relative to North and varies between 0 and 360°. Strike determines the direction of fault which is defined as the line of intersection of the fault plane and the ground surface. The strike of a fault is measured in such a way that the hanging wall is always kept on the right and footwall block on the left. Rake angle λ shows the direction of relative motion of hanging wall with respect to footwall. It is measured relative to fault strike and it varies between $\pm 180°$.

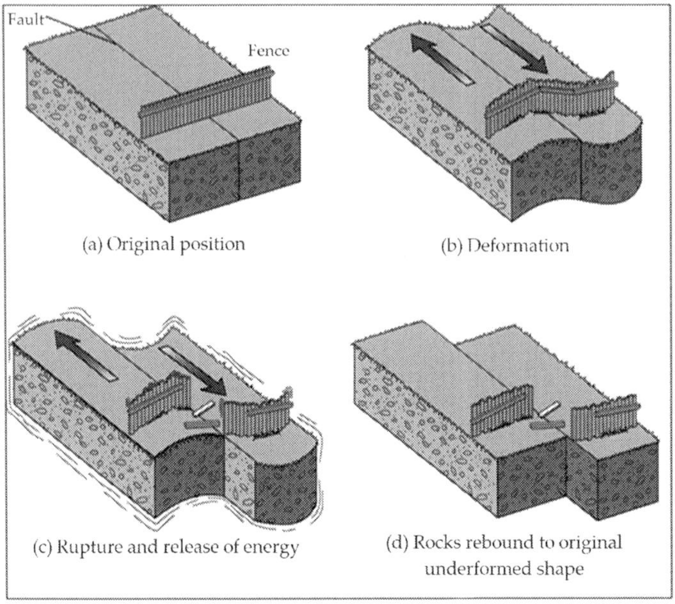

Figure 1.13: Concept of Elastic Rebound Theory

Based on geometrical properties discussed above and on type of movement, faults can be divided into various types. Movements in two blocks of faults can be horizontal, vertical and both. When blocks movement are horizontal in nature relative to each other, the fault is called strike-slip fault. If the hanging wall in strike-slip fault moves away from a person standing on the fault plane facing towards the direction of the strike, it is called left-lateral fault. Value of Rake angle for left-lateral fault (λ) is zero. If the direction of the horizontal movement is reversed i.e. hanging wall moves toward the observer, then it is called right-lateral fault. Value of rake angle is $\pm 180^{\circ}$ in this case.

In case of vertical movements of the two blocks along the fault plane, faults are termed as dip-slip faults. Upwards motion of hanging wall with respect to footwall results in formation of normal dip-slip fault. Rake angle (λ) is greater than zero in this situation. Reverse faults occur if direction of hanging wall and footwall are reversed. When rake angle has very small values ($0< \lambda<30^{\circ}$), the fault is termed as thrust fault. Generally, movements of two blocks along the fault plane are both horizontal and vertical. Faults are termed as oblique faults when movements of hanging wall are both horizontal and vertical. They are described on the basis of dominant slip direction. All these movements are shown in Fig.1.13. In case two tectonic plates converge, reverse dip-slip faults occur. Normal faults are the result of diverging tectonic plates. Strike-slip faults typically exist in transform boundaries (Sucuoğlu & Akkar, 2014).

SEISMIC WAVES

Strain energy is released due to sudden rupture of a fault propagates in form of seismic waves. There are two types of seismic waves, P-waves and S-waves. P-waves are called compression waves and S-waves are shear waves. P-waves have higher speeds than S-waves and are hence, observed first on the seismogram. Equations 1.1 and 1.2 describe the velocities of P-waves and S-waves respectively. Their velocities depend on the type of medium they travel.

$$V_p = \sqrt{\frac{(1-\upsilon)}{\rho(1+\upsilon)(1-2\upsilon)}} \qquad (1.1)$$

$$V_s = \sqrt{\frac{E}{2\rho(1+\upsilon)}} = \sqrt{\frac{G}{\rho}} \qquad (1.2)$$

E and are the modulus of elasticity and mass density of the elastic medium, respectively. υ represents Poisson's ratio (0.25) and G is the shear modulus in above equations. P and S-wave velocities increase with depth as values of E and G increase towards the interior part of the earth's crust. Usually, P and S-wave have velocities of 6 km/s and 4 km/s respectively within the crust. In general, P-waves are expected to be times faster than S-waves.

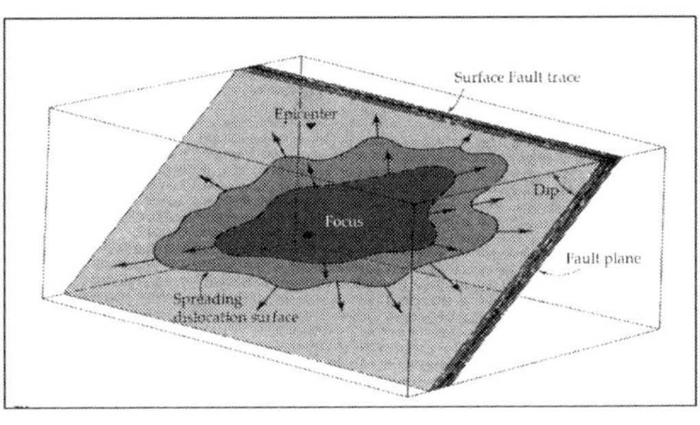

Figure 1.14: Earthquake Fault Zone

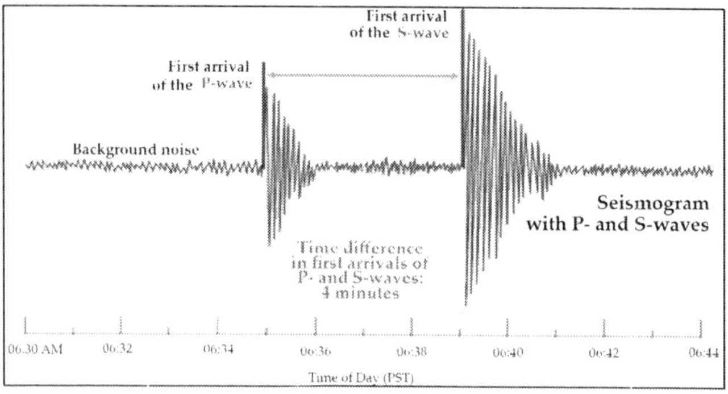

Figure 1.15: A Conceptual Diagram of S-P lag-time

P-waves propagate through the medium by moving particles in the direction of the travel. Hence, P-waves are called longitudinal waves. S-waves, on the other hand, while traveling through a medium, move the particles perpendicular to the direction of travel and hence, are called shear waves. Due to this fact, S-waves cannot travel through liquid mediums (e.g., outer core). S-waves are further divided into SH and SV waves according to the particle motions in the horizontal and vertical planes, respectively. The particles move in horizontal plane in case of SH waves generating lateral shaking that might result in large dynamic deformations of structures. As P-and S-waves are generated immediately after the fault rupture and propagate through the solid body of the Earth's crust, they are commonly known as body waves.

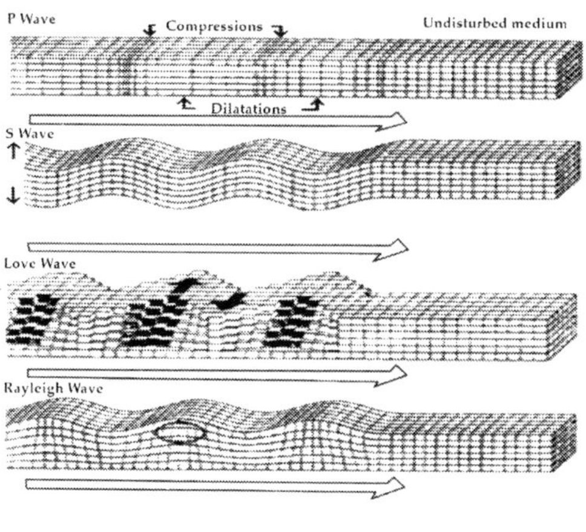

Figure 1.16: Types of Seismic Waves

When these body waves travel across the Earth's surface, they are called surface waves. The surface waves do not travel towards the interior of the Earth. Hence, their amplitude decreases as depth increases. Surface waves are divided into Love (LQ) and Rayleigh (LR) waves. Love waves are trapped SH waves that propagate along a horizontal layer between the free surface and the underlying elastic layer. Trapped SH waves travel across by reflecting from the top and bottom of the horizontal layer. The velocity of the Love waves lies between the shear-wave velocities of the horizontal layer and the underlying half space. The particle motion of propagating Rayleigh waves is polarized in the vertical plane due to trapped P and SV waves. The velocity of Rayleigh waves is approximately 90 % of the shear-wave velocity of the elastic medium if the Poisson's ratio υ is assumed as 0.25. Since surface waves are trapped within a boundary, they can travel long distances along the Earth's surface. Their wave lengths and periods are longer. Their propagation velocities depend on the elastic properties of the medium and their periods. The method of propagation along with different particle movements for P, S, Love and Rayleigh waves is shown in Fig.1.16.

MAGNITUDE OF AN EARTHQUAKE

Magnitude scales measure the size and energy released by earthquakes. The first magnitude scale was proposed by Richter in 1935 for quantifying the sizes of earthquakes in southern California from the maximum amplitudes (A in mm) of seismograms recorded by the Wood-Anderson seismographs. The local magnitude (M_L) expression proposed by Richter is given in Equation (1.3).

$$M_L = log(A) - log(A_0) \quad (1.3)$$

The above equation calibrates M_L with base amplitude A_0. M_L corresponds to the amplitude of a base earthquake that would yield a maximum trace amplitude of 0.001 mm on a Wood-Anderson seismograph located at an epi-central distance of 100 km. Richter provided the calibration factor "-log (A_0)" for epi-central distances up to 1000 km for the average conditions in southern California. Values for M_L can also be obtained from the nomogram shown in Fig. 1.17. In order to use nomogram, arrival times of P- and S-wave are required along with the maximum amplitude readings on a Wood-Anderson seismograph. The calibration by base amplitude A_0 is embedded into the nomogram. If the time difference between P- and S-wave arrivals is 25 s and the maximum amplitude of the Wood-Anderson seismogram is 20 mm, nomogram graphically estimates the value of 5 for M_L.

Richter scale calibrations are only valid for southern California. In order to use this scale in any other part of the world, the difference in regional depletions of seismic waves should be considered while doing calibrations. A faulty estimation for magnitude of an earthquake in regions other than California might be done if regional factors are not considered properly. Other scales will be discussed in later sections of this book.

INTENSITY OF AN EARTHQUAKE

There can be two kinds of measurements which could explain the intensity of earthquakes. These measurements can be of qualitative and quantitative type. For quantitative measurements, different seismic instruments are used. Subjective personal observations describe the intensity of an earthquake in a qualitative manner. Qualitative measurements are done by seismic engineers and scientists using predefined indices. Instrumental observations are the most reliable presentation of measurement of an earthquake. Both instrumental and observational intensities are discussed in following sections.

Instrumental Intensity

To describe the earthquake intensity, accelerograph provides the most reliable and useful data. The recordings made by accelerograph are termed as accelerograms or accelerometric data. It records the acceleration of the particles under ground shaking and these recordings are time dependent. Accelerographs are usually installed in the field in seismically active areas to record the strong vibrations caused by the earthquake. Normally, they record mutually perpendicular components of motion i.e. vertical and two horizontal directions. Accelerographs are either analogue or digital. The analogue accelerographs are the first-generation instruments and they record on film papers (Fig. 1.17).

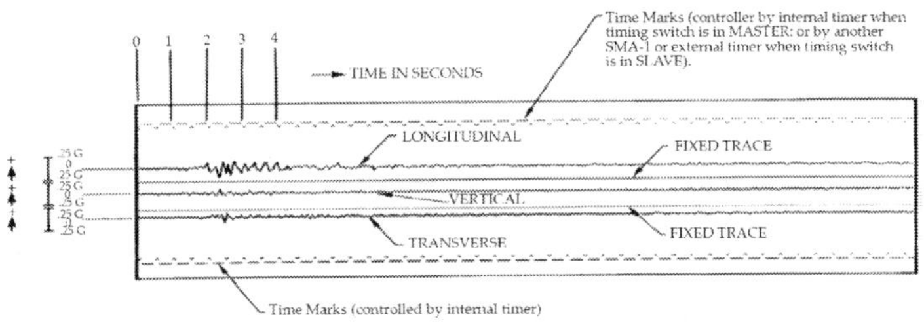

Figure 1.17: Simulated Earthquake Recording Showing Polarity Conventions

A threshold acceleration level is required for triggering the accelerograph to record the data. This triggering condition for activating the recording of data may fail to record the initial seismic waves if these are below the threshold level. The computation of ground velocity of seismic waves may be affected due to this fault. Another limitation of analogue accelerograph is the amount and quality of data it records as the data is recorded on a film paper.

Digital accelerographs were introduced to record the data almost 50 years after the first analogue accelerograph had been used. They operate continuously and have pre-event memory. Pre-event memory for these accelerographs is limited to 15 seconds. This means that they start recording 15 seconds earlier than the waveforms start arriving at the station. Due to this property, they can record the first wave arrivals and the data obtained is more reliable than the analogue version.

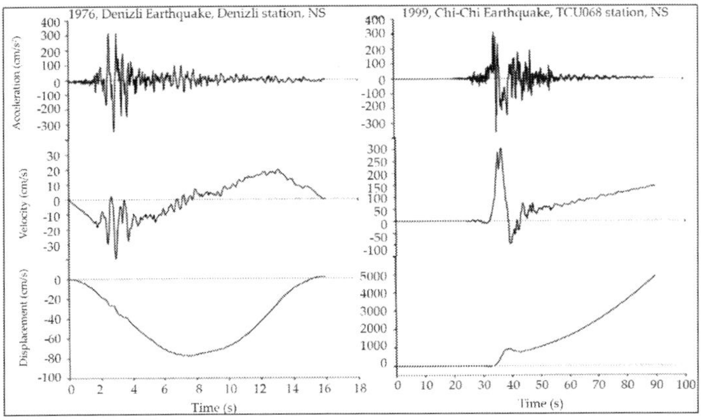

Figure 1.18: Accelerograms demonstration for Denizli and Chi-Chi Earthquakes

Accelerograms provide significant information about the nature of ground earthquake and also about the highly varied characteristics that are different for each earthquake or within an earthquake at different locations. Ground-motion parameters such as peak ground acceleration, velocity or spectral ordinates that are obtained from the accelerograms quantitatively describe the intensity of an earthquake.

Observational Intensity

Macro seismic intensity is a qualitative measurement of the influence of ground movements in a particular area resulted from an earthquake. It consists of an index which defines the strength of ground trembling based on observed in a particular area. Though in presence of highly sensitive instruments recording the actual intensity, use of observational qualitative measurements seems less significant. But they are essential in assessing effects of earthquakes on structure vulnerability and for the revision of historical seismic activity that is used in seismic hazard assessment.

The first internationally used intensity scale was developed by Rossi and Farel in 1883. It was called Rossi-Farel scale and it was a 10-degree scale. Seiberg's 12-degree scale constitutes the basis of modern intensity scales and was proposed in 1923. Its later version is known as Mercalli-Cancani-Sieberg scale (MCS) and was introduced in 1932. The older version was translated into English by Wood and Neumann in 1931 and was named Modified Mercalli (MM) scale. Modified Mercalli scale was entirely revised by Richter in 1956. Medvedev, Sponheuer and Karnik published MSK scale in 1964 based on MCS scale and was widely used in Europe before the publication of European Macroseismic Scale (EMS) in 1998 (Grunthal 1998).

An EMS intensity scale is now globally used in order to measure the intensity of earthquakes in qualitative manner. Table 1.1 describes the EMS intensity scales. Different intensities are described by considering effects on humans, natural and manmade objects including buildings. Buildings are studied in terms of damage caused to them due to seismic activity. EMS scale also involves six vulnerability classes based on construction type, material used and structural conditions. These six classes are represented in Fig.1.19. The damage is described as structural and non-structural considering five damage grades: negligible-to-slight, moderate, substantial-to-heavy, very heavy and destruction. These grades are described verbally and visually in EMS. For masonry and reinforced concrete buildings, these damage grades are represented in Fig. 1.19.

Table 1.1: EMS Intensity Vs Level of damage outcome after Earthquake

EMS Intensity	Definition
Not Felt	(a) Not felt, even under the most favourable circumstances (b) No effect (c) No damage
Scarcely Felt	(a) The tremor is felt only in isolated instances (<1 %) of individuals at rest and in an especially receptive position indoors (b) No effect (c) No damage
Weak	(a) The earthquake is felt indoors by a few. People at rest feel a swaying or light trembling (b) Hanging objects swing slightly (c) No damage
Largely Observed	(a) The earthquake is felt indoors by many and felt outdoors only by very few. A few people are awakened. The level of vibration is not frightening. The vibration is moderate. Observers feel a slight trembling or swaying of the building, room or bed, chair etc. (b) China, glasses, windows and doors rattle. Hanging objects swing. Light furniture shakes visibly in a few cases. Woodwork creaks in a few cases. (c) No damage
Strong	(a) The earthquake is felt indoors by most, outdoors by few. A few people are frightened and run outdoors. Many sleeping people awake. Observers feel a strong shaking or rocking of the whole building, room or furniture. (b) Hanging objects swing considerably. China and glasses clatter together. Small, top-heavy and/or precariously supported objects may be shifted or fall down. Doors and windows swing open or shut. In a few cases window panes break. Liquids oscillate and may spill from well-filled containers. Animals indoors may become uneasy. (c) Damage of grade 1 to a few buildings of vulnerability class A and B

Slightly Damaging	(a) Felt by most indoors and by many outdoors. A few persons lose their balance. Many people are frightened and run outdoors. (b) Small objects of ordinary stability may fall and furniture may be shifted. In few instances dishes and glassware may break. Farm animals (even outdoors) may be frightened. (c) Damage of grade 1 is sustained by many buildings of vulnerability class A and B; a few of class A and B suffer damage of grade 2; a few of class C suffer damage of grade 1.
Damaging	(a) Most people are frightened and try to run outdoors. Many find it difficult to stand, especially on upper floors (b) Furniture is shifted and top-heavy furniture may be overturned. Objects fall from shelves in large numbers. Water splashes from containers, tanks and pools. (c) Many buildings of vulnerability class A suffer damage of grade 3; a few of grade 4. Many buildings of vulnerability class B suffer damage of grade 2; a few of grade 3. A few buildings of vulnerability class C sustain damage of grade 2. A few buildings of vulnerability class D sustain damage of grade 1.
Highly Damaging	(a) Many people find it difficult to stand, even outdoors (b) Furniture may be overturned. Objects like TV sets, typewriters etc. fall to the ground. Tombstones may occasionally be displaced, twisted or overturned. Waves may be seen on very soft ground. (c) Many buildings of vulnerability class A suffer damage of grade 4; a few of grade 5. Many buildings of vulnerability class B suffer damage of grade 3; a few of grade 4. Many buildings of vulnerability class C suffer damage of grade 2; a few of grade 3. A few buildings of vulnerability class D sustain damage of grade 2.
Destructive	(a) General panic. People may be forcibly thrown to the ground (b) Many monuments and columns fall or are twisted. Waves are seen on soft ground. (c) Many buildings of vulnerability class A sustain damage of grade 5. Many buildings of vulnerability class B suffer damage of grade 4; a few of grade 5. Many buildings of vulnerability class C suffer damage of grade 3; a few of grade 4. Many buildings of vulnerability class D suffer damage of grade 2; a few of grade 3. A few buildings of vulnerability class E sustain damage of grade 2.

Very Destructive	(a) Most buildings of vulnerability class A sustain damage of grade 5. Many buildings of vulnerability class B sustain damage of grade 5. Many buildings of vulnerability class C suffer damage of grade 4; a few of grade 5. Many buildings of vulnerability class D suffer damage of grade 3; a few of grade 4. Many buildings of vulnerability class E suffer damage of grade 2; a few of grade 3. A few buildings of vulnerability class F sustain damage of grade 2.
Devastating	(a) Most buildings of vulnerability class B sustain damage of grade 5. Most buildings of vulnerability class C suffer damage of grade 4; many of grade 5. Many buildings of vulnerability class D suffer damage of grade 4; a few of grade 5. Many buildings of vulnerability class E suffer damage of grade 3; a few of grade 4. Many buildings of vulnerability class F suffer damage of grade 2; a few of grade 3.
Completely Devastating	(a) All buildings of vulnerability class A, B and practically all of vulnerability class C are destroyed. Most buildings of vulnerability class D, E and F are destroyed. The earthquake effects have reached the maximum conceivable effects.

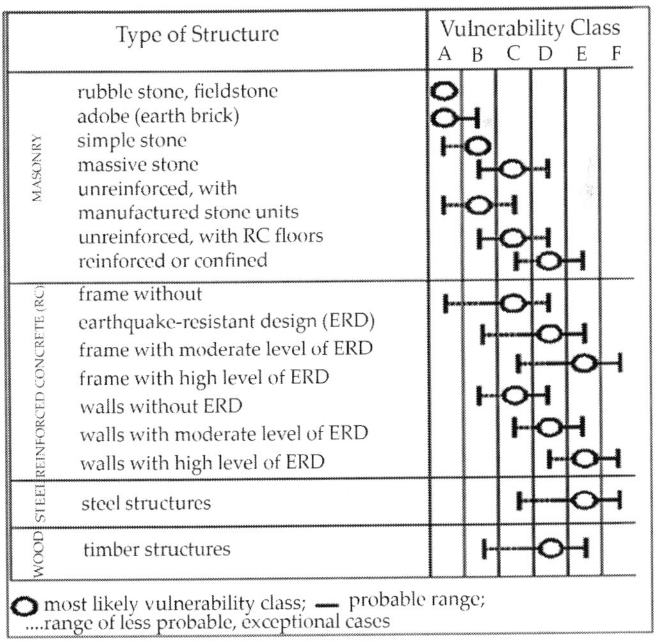

Figure 1.19: Vulnerability Class vs. Types of Structures

As intensities in EMS scale are highly influenced by various factors like hydrological conditions in the earthquake area etc., it is a possibility that sometimes, the direct effects of earthquake shaking might not be recorded properly. EMS scale does not consider the effects of ground failures such as landslides, cracks in the ground etc. in intensities rather associates them with different categories. As lower intensities are mainly described by human reactions, intensities of engineering concerns are usually VII and above. X and XI level intensities can be defined as the upper limit while intensity level XII represents a very unlikely strong ground shaking as shown in table 1.1. Classification of damages has been also elaborated in table 1.2.

Table.1.2: Classification of Damage to Masonry and Reinforced Buildings

Masonry buildings	Reinforced buildings	Classification of damages
	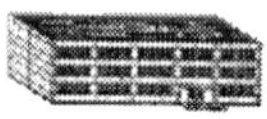	Grade 1: Negligible to slight damage (no structural damage, slight non-structural damage)
	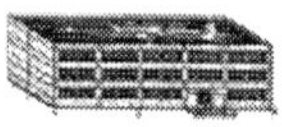	Grade 2: Moderate damage (slight structural damage, moderate non-structural damage)
		Grade 3: Substantial to heavy damage (moderate structural damage, heavy non-structural damage)
	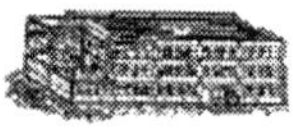	Grade 4: Very heavy damage (heavy structural damage, very heavy non-structural damage)
		Grade 5: Destruction (very heavy structural damage)

The intensity data is drawn on the map in order to study the effects of earthquake in the area. This map is known as isoseismal map. Intensity points are drawn as contours.

Isoseismal contours represent the areas with high enough earthquake intensity to cause damage. Close isoseismal contours represent the area with fault rupture and the epicentre would be expected to lie within or very close to the highest contour. The distribution of the isoseismals, and the distances between them reflect primarily the magnitude and the focal depth of the earthquake. The isoseismals of shallow earthquakes tend to be more closely spaced, whereas for deeper earthquakes, in subduction zones, they will generally produce lower values of intensity but isoseismals will cover much larger areas. An example of isoseismal map is shown in Figure 1.20 obtained after the main shock of the 1976 Fruili earthquake in Italy (Goretti & Di Pasquale, 2004; Mucciarelli et al., 2004).

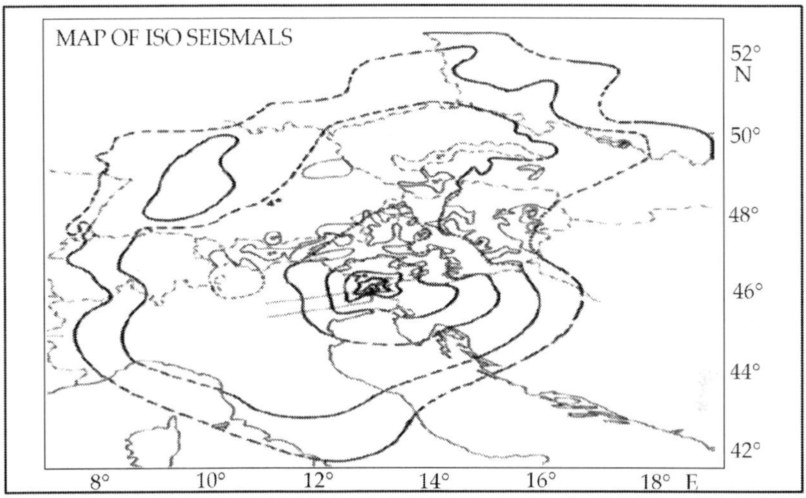

Figure 1.20: Map of Isoseismals

EFFECTS OF EARTHQUAKES ON BUILT ENVIRONMENT

Effects of earthquakes can be divided into two categories: direct and indirect. Strong ground vibrations below the foundation of a structure is an example of direct effect. A fault fracture across a long structure such as pipe lines, viaducts or road networks are also considered to be direct effects. Geotechnical failures such as landslides, liquefaction, tsunami and fire resulting from an earthquake are indirect effects. These indirect effects directly influence the structures located in those areas. Indirect effects are also very important and can cause lots of damage. Death tolls and economic damage for tsunamis resulted from the earthquakes of 2004 in Indonesia and 2011 in Japan were far greater than those from the direct effects. Strong ground vibrations, fault rupture and geotechnical deformations are some of the factors which are discussed in this section. In general, these factors have the highest impact on the structures and hence, are very important to be considered.

Strong Ground Vibrations

All structures resist the vibrations or movements caused due to earthquake. The resistive forces are in fact the inertial forces due to their mass as explained by Newton's law of motion. When the ground movements increase beyond the inertial resistive forces, structures are deformed beyond their linear elastic deformation range and hence, damage caused is permanent. A structure might collapse if it does not have sufficient elastic deformation capacity. Collapse of a building usually results in fatalities.

Fault Rupture

Large horizontal and vertical deformations are produced due to crossing of a surface fault rupture through a structure. Probability of fault rupture crossing a large structure such as highways, railway lines or a pipeline system is high. Damage in case of railway is repairable as railway bends are flexible and can be replaced easily. The case with highways and pipelines is the same but still economic losses might be considerable in some cases (i.e. an underground water pipeline gets damaged). However, if a fault crosses a stiffer structure such as buildings, bridges, dams etc., the damage is not easily repairable and it might have effects on other structures in the vicinity of these damaged structures.

Geotechnical Deformations

Settlements occur in non-compacted soil deposits due to strong ground vibrations. Deformations in highways and railways constructed over such non-compacted soil layers are typical examples of such ground deformations that have resulted in severe earthquakes. These settlements may be uniform, tilting or differential in nature. Uniform and tilting settlements usually do not result in cracks in buildings but differential settlements result in cracks. Large amounts of differential settlements can cause partial collapse of a building in some cases or make the building inhabitable, which is shown in Fig.1.21.

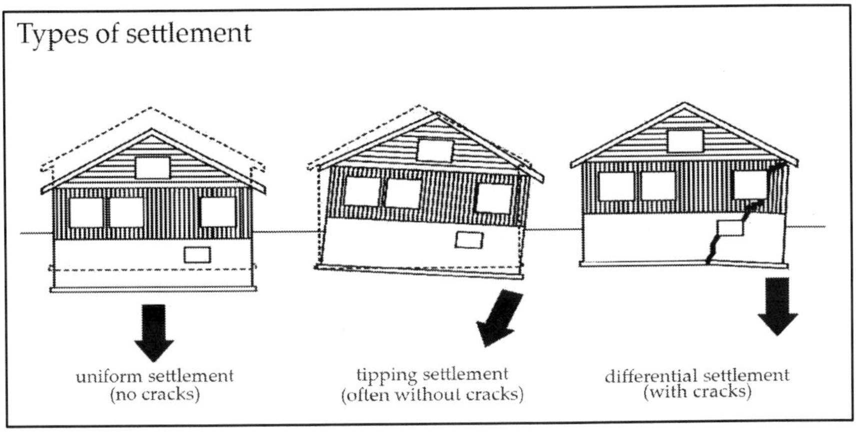

Figure 1.21: Foundation Failures by Settlement type

Sometimes, saturated soil deposits lose their strength under cyclic stresses caused by earthquake and start flowing as a plastic material. This phenomenon is called liquefaction. Large vertical deformations result when bearing capacity of the liquefied soil becomes less than the static gravity stresses applied by the structures above. In this case, structures usually sink into the soil and tilt with a rigid body motion during the earthquake. Ground accelerations in liquefied soils are usually very low and do not cause considerable damage structural damage.

Hence, liquefaction may not cause life losses or injuries, but it can cause significant economic losses due to tilted or displaced structures. In some cases, extensive ground displacements may lead to partial collapse, loss of stability and significant damage. Figure 1.22 shows buckled houses in the Marina District after the 1989, Loma Prieta earthquake.

Lateral spreading is the lateral deformation in soil slopes when subjected to earthquake vibrations. Laterally spreading liquefied soils can drag the structures above them, sometimes into the sea at a shoreline. Values of lateral spreading may be estimated using predictive methods, simplified analytical models, or numerical (finite elements, finite difference, etc.) methods shown in Fig.1.23. Precision and accuracy are expected to increase from predictive methods to finite element methods along with increase in time and effort.

Figure 1.22: Destruction of Buildings due to Earthquake

Soil slopes that are stable under static conditions might lose their stability during earthquakes due to additional dynamic stresses. As a result, an earthquake induced landslide is occurred. Current evaluation of seismic slope performance is through irreversible (or residual) slope displacements. Similar to the evaluation of lateral spreading, irreversible slope displacements could be estimated by means of predictive frameworks, simplified analytical models or numerical methods(Sucuoğlu & Akkar, 2014).

Figure 1.23: Deformation of Soil, especially on Shore lines

2
CHAPTER

ENGINEERING SEISMOLOGY

INTRODUCTION

Seismology can be defined as the scientific study of earthquakes and the propagation of elastic waves through the Earth. Earthquake environmental effects tsunamis, volcanic, tectonic and other sources of earthquakes are also included in seismology.

The catastrophic nature of earthquakes makes them the most probable causes of casualties and damages to populated areas. Though their location is not the same, the way of their occurrence is same. They occur without warning and turn cities into scrap and kill large number of people. During the 20th century, ten major earthquakes with death toll more than 50,000 people and over 100 earthquakes with death toll more than 1000 people have occurred across the world (FEMA 383, 2003). The Tangshan earthquake of 1976 in China was one of the deadliest earthquakes killing more than 600,000 people. The magnitude of this enormous earthquake was an 8 on Richter scale. In 1994, an earthquake with a magnitude of M 6.7 struck Los Angeles (located in Northridge) killing 60 people. Compared to the Tangshan earthquake, this was not devastating at large scale, but this changed the direction of earthquake engineering research around the world. Another earthquake in 1995 struck in Kobe-Japan with magnitude M 6.9 with 5600 fatalities. This earthquake also played an important role in giving new direction to the earthquake research. Among various reasons, two major reasons resulted in this change in direction of research.

The two above-mentioned earthquakes, one in USA and Japan each, though did not kill many people, they had resulted in enormous property damage. In other words, the economic losses were immense as both locations were densely populated. This was the first reason which initiated the change of direction in earthquake research. In USA, around 15,000 buildings were demolished due to their failure in serviceability resulting in a total loss between $15 billion to $40 billion. The repair costs for the 180,000 buildings effected from earthquake in Japan was estimated between $90 to $150 billion. Due to such huge economic losses, it was realized that the old principles for seismic designs need to be changed. Previously, major goal of design principles was to protect buildings from collapse. Principles of performance based design were set up to reduce the damage costs by keeping the damage in non-structural and structural members up to an acceptable level.

Because both above mentioned earthquakes, despite of being moderately strong in magnitude, caused lots of damage to the infrastructure and buildings, it was noted that epicentres of both earthquakes were located near densely populated areas. This proves that the location of an earthquake's epicentre is directly related to the amount of losses. If both of these earthquakes are supposed to happen today in same areas, estimated losses will range between $100 to $200 billion and over $1000 billion for USA and Japan respectively (Victor Gioncu & Mazzolani, 2010). This shows that even earthquakes of moderate magnitude can be catastrophic if their epicentre is located near urban areas. As with increase in population, urban areas close to epicentre locations are also growing. This reason became the second most important to study the effects of near source earthquakes on the structures built in epicentral areas.

Structural responses can be predicted with detailed accuracy but they will not be effective if the nature of seismic actions and movements is not determined effectively (Gioncu and Mazzolani, 2006). So, it is very necessary to bridge the gap between engineering seismology and earthquake engineering in order to achieve a safe, effective and efficient seismic design.

DIFFERENT CONSIDERATIONS FOR SEISMIC DESIGN

Among various other objectives of seismic design, advancement in understanding earthquakes and their effects is the most important one. Various disciplines of Seismology, Engineering Seismology and Earthquake Engineering overlap to achieve this objective. In order to achieve a safe seismic design, consideration of different aspects of above mentioned fields is very important. Some of these are described briefly in following paragraphs (FEMA 383, 2003).

Improve earthquake monitoring

Identification of seismic hazards and their risk assessment are the most critical components of earthquake relief strategy. Therefore, a monitoring system has been developed consisting of regional networks of stations equipped with different earthquake recording instruments. These stations are selected using satellite based observations (GPS monitoring) and data centres are associated with these stations (FEMA 383, 2003)(Holzer, Scawthorn, & Rojahn, 2005). Accelerations, velocities or displacements are considered to be most useful variables for seismic design against which data is obtained from these stations. Specifying locations of these stations on a regional earthquake monitoring network is very difficult. This is because site and spacing requirements for seismic research are different than for earthquake research. As seismologist are focused towards the study of earth's structure, they want their stations to be located in quiet areas and away from human activities as much as possible. In case of earthquake engineers, stations near built environments of the urban areas are preferred. Strong movements are considered useful for engineering purposes if they are recorded with in 20 to 50 km from the point of earthquake-rupture. The amount of distance can vary depending upon the soil conditions. Hence, a regional seismic network of stations cannot be used for earthquake engineering purposes. This makes it quite expensive for areas which are lightly populated as dense station network is impractical and in some cases, impossible. So, a compromise between these networks has to be made in such a way that data obtained is useful for both purposes i.e. seismological and earthquake engineering(Lee, Jennings, Kisslinger, & Kanamori, 2002).

Improve understanding of earthquake occurrence:

Along with other fields, seismology has also advanced significantly since past few decades. Modern technologies such as Global Positioning Systems (GPS) has helped seismologists to predict overall seismic activity in long terms. Still, earthquakes pose greatest threats to humankind. This is evident from the devastating earthquakes occurred during last century and beginning of this one. The most difficult part of forecasting the occurrence of earthquakes is that earthquake's occurrence involves lots of elementary processes. Even if the mechanisms behind these elementary processes is understood completely, their interaction between themselves is so complex that the earthquake forecasting becomes really difficult (Kanamori, 2005). The initial understanding of earthquakes is developed by plate tectonics and continental drift theories. These theories are described briefly in Chapter 1. As plate boundaries are in constant motion, ninety percent of the seismic activity takes place along the plate boundaries. Still, the type of earthquakes depends upon the nature of the interaction between two plates. Remaining ten percent of the seismic activity occurs away from current plate boundaries. Some seismologists consider these intra-plate earthquakes to be occurring along the ancient plate boundaries. As described in the previous Chapter, the physical characteristics of intra plate earthquakes are different from those of inter plate earthquakes. Characteristics of both inter and intra – plate earthquakes should be understood in detail in order to make future predictions.

Improve fundamental knowledge of earthquake effects:

In order to reduce the losses resulting from the earthquakes, it is necessary to improve the knowledge regarding the earthquake effects in the past and potential damages that can be caused by them at present. Modeling the earthquake effects such as the source properties, wave propagation phenomenon from source to the site and local site conditions. This procedure generates synthetic seismograms for the expected future earthquakes which improves the understanding of rupture process, gives information about the nature of the fault and the properties of earth's crust in the surrounding area. Different parameters such as ground accelerations, velocities and displacements, vibration's duration etc. which cause soil liquefaction, landslides and structural damages can be estimated as well. These parameters should be simulated accurately as they are considered by structural engineers during the seismic design process.

Improve the seismic design of structures:

With advancements in computer technologies, computer based simulations are being used along with integrated experiments, theory and databases in order to understand the behaviour of different structural components of a structure. This is one of the new features of earthquake engineering research. In order to establish new methodologies for performance based designs, different types of earthquakes and the resulting behaviour of structures due to these earthquakes is being studied in detail. Consideration of different design philosophies for structures situated in low to moderate and strong seismic areas is very important as it will become uneconomical if structures are over designed and unsafe if structures are under – designed. Objectives of new technologies also include development of new materials, structural systems and technologies of earthquake resistant structures.

Start development of next generation performance-based codes:

The objective of all above procedures is to achieve a safe and economical seismic design by reducing the structures vulnerability towards seismic activity. Learning from the effects of previous strong earthquakes and significant developments in knowledge regarding seismic phenomena, effort is required to bring all this theoretical knowledge towards practical implementation. This includes upgrading design guidelines and developing new codes, promoting education for practicing professional and other methods. This will help in achieving the target of better earthquake resistant structures.

DISCIPLINES INVOLVED IN SEISMIC DESIGN

As discussed in previous section, different disciplines are involved in the process of seismic design. The process of seismic design along with relationships of other disciplines is presented in Figure.

The most important objective of Engineering Seismology is to use the seismological knowledge for making design for buildings safe against earthquake effects. This is achieved by considering the characteristics of source, site and nature of movements resulted from a particular earthquake. Earthquake engineering, on the other hand, focuses on solving the problems of construction vulnerability. Earthquake engineering is a branch of more generalized field, structural engineering whose purposes include designing safe structures by studying the nature and magnitude of different forces which a structure is expected to bear in future such as dead, live, snow load etc.

Data obtained from engineering seismology is incorporated in structural analysis and a complex examination of structures is done along with an engineering overview of the designed structure. The final design obtained is seismic design whose objective is to reduce the structural damage in case of a strong earthquake thus reducing economic losses. A detailed overview of relation of all the disciplines is discussed in the following section.

Seismology and Engineering Seismology

Seismology involves study of tectonic plate movements which are associated with occurrence of earthquakes. Seismologists are specialized in geophysics and being earth scientists, the focus on analysis of production and propagation of seismic waves through geological materials. This includes studying variety of different topics such as relation between faults stress and seismicity, interpretation of rupture mechanisms from obtained seismic wave's data, integration of geoscientific information for defining seismic zones and collaboration with engineers to minimize the damages to constructed structures.

Concepts and methodologies used in seismology are much different from those used by engineers. They can propose, test and then even reject the false hypotheses. For example, generation of earthquakes was explained using different theories such as chemical reactions taking place in earth's interior, continental drift or expansion of earth, all these types of incorrect hypotheses were used without any consequences. As the causes of earthquakes have been agreed upon and Plate Tectonic theory is accepted by the entire scientific community, seismologists are focusing on understanding the types of faults and types of movements to progress in field of qualitative explanation of the earthquake phenomenon.

Structural engineers, on the other hand, are more involved in design of buildings and other civil structures. They rely on proven principles and data as they cannot rely on testing their false hypotheses as it can cause large human and economic damages. They require quantitative data and explanations in order to consider any kind of factor, even in case of earthquakes. Forces generated due to earthquakes must be known accurately so that structure can be designed to have resistance greater than these forces in order to avoid failure.

It should be noted that in order to achieve a safe seismic design, interests of seismologists and structural engineers frequently overlap, but not in the same way. Both are though seeking knowledge regarding active faults locations. Seismologists want to know about their movements whereas engineers want to know the potential for the future earthquakes and building behaviour resulted from the earthquakes and then use this information for structural design. In other words, it can be said that seismologists use more theoretical approach or develop theories whereas engineers try to implement those theories practically. Implementation of these theories in practical way is very important which can be done by transferring theoretical knowledge from seismology to practical, structural engineering. Now, as the context has been clarified, engineering seismology has a particular goal of transforming the qualitative knowledge in some quantitative form so that it could be used by structural engineers (Hu, 2001).

Tasks of Seismology

One of the main objectives of Seismology to study and elaborate theories concerned with generation of earthquakes and travelling mechanisms of seismic waves. Simultaneously, it is also involved in recording of seismograms and their interpretation. One of main aspects of Seismology is source characterization and this involves:

- Confirmation of Tectonic plates along with their boundaries.
- Determination of tectonic plate movements and establishment of stressed zones with help of GPS stations.
- Development of better understanding towards nature of friction and rupture in fault zones.
- Establishment of basic fault types.
- Discovery and documentation of ancient and near past earthquakes.
- Recording and processing of recorded ground motions using seismic stations.
- Determination of earthquake strength, frequency and analysis of seismic hazard.
- Publication of collected data so that it can be used for other purposes and by other fields.
- Publication of seismic maps with inclusion of information regarding known faults, seismic areas and their features.
- Prediction of future earthquakes.

Tasks of Engineering Seismology

Tasks of engineering seismology include

- Analysing the influence of seismic waves along the travelled path and their depletion trends.
- Identification of different site conditions and their behaviour towards seismic activity.
- Identifying nature and properties of movements i.e. direction, etc.
- Studying effects of different topographic areas and alluvial valleys on nature of ground movements.
- Studying soil liquefaction and its effects.
- Classification of different types of ground motion.
- Conversion of recorded data such as accelerations, velocities etc. into digital form.
- Estimating the duration of ground motions.
- Determining number of significant pulses.

- Properties and differentiation of near and far-field earthquakes

Among primary tasks of engineering seismology, collection of general information about the seismic region is the basic one. These tasks are accomplished through following activities.

- Data collection about historical earthquake occurrences including their location.
- Investigation of tectonic movements in high seismic areas.
- Forecasting magnitudes of strong vibration resulted from an earthquake with moderate strength in a seismic area.
- Discovering and studying dormant faults which can become active using modern technologies.

After collecting general information about the seismic area, the next task is to collect information about properties of earthquake source of generation. This includes investigation of following aspects.

- Type of earthquake. This is done by consideration of tectonic plate boundaries for inter-plate and existing plate faults for intra-plate earthquakes.
- Depth of focus point and differentiating surface and deep sources.
- Defining the tendencies of plate movements and inferring the direction of fault ruptures.
- Geographical properties of ground motions.

The next task is about gathering information regarding geotechnical load conditions. This is related to study earth condition in the area. Following aspects are studied to accomplish this task.

- Thickness of soil layer and their horizontal and vertical structure along with their dynamic properties.
- Study of Hydrological features of the region regarding their role in augmentation of ground movements.
- Probability of occurrence of ground liquefaction or landslides.

Final tasks of engineering seismology are

- Presenting accelerograms.
- Maximum possible value for earthquake magnitude.
- Magnitude frequency.

- Earthquake duration.
- Decreasing trends of seismic waves in relation with distance from epicentre.

Because of difficulties in obtaining details about crust and mental properties and their transformation to quantitative values, information provided by engineering seismology is highly uncertain. Statistical procedures are needed to be applied on the data to estimate these quantitative values because data might differ for two earthquakes occurring due to approximately same source.

EARTHQUAKE AND STRUCTURAL ENGINEERING

Earthquake Loads versus Static Loads

As it is quite difficult to define the nature of earthquake loads, it is easier to compare them with other conventional type of loading such as wind loads. Both wind and earthquake loads are horizontal but their application is different physically. Their comparison is shown in figure. While designing for wind loads, entire area that is exposed to wind is taken into consideration. This is not the case with earthquake loads. For earthquake loads, inertia forces due to ground movements are considered while analysing the structure strength for earthquake loading. This is shown in figure a.

The oscillating nature of wind load is very minimum. There may be very small fluctuations in wind direction but stress reversal happens only when direction of wind is reversed. Usually, this takes place after a long period, sometimes even hours. But in case of earthquakes, stress reversal occurs in matter of seconds. This is due to the cyclic nature of earthquake movements. This is explained in figure b.

Probability of wind loads exceeding their limit throughout the life span of a structure is higher than that for earthquake loads. Wind loads may reach their maximum limit several times during life time of a structure whereas earthquake loads may reach once or twice the maximum value during structure's life. Figure c represents the probability of occurrence of maximum value for wind and earthquake loads. The time period for wind loads is calculated in days whereas for earthquake loads, it is measured in years. Based on these differences, design philosophy for earthquake loads is much different than that for wind loads.

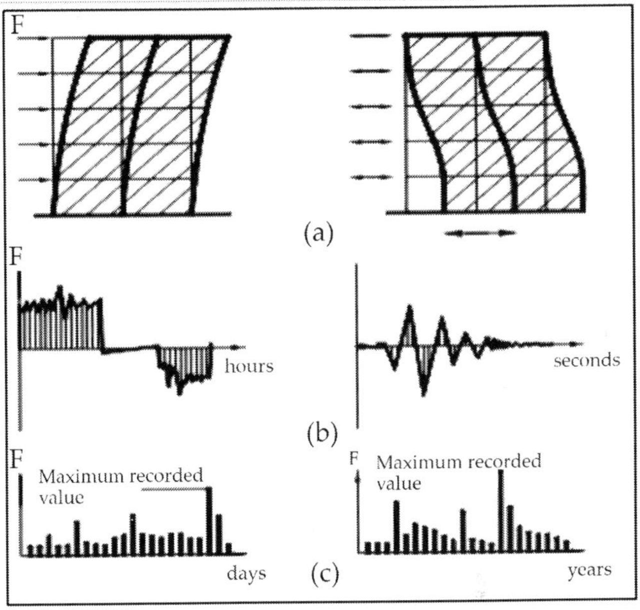

Figure 2.1: Illustration of structure under Seismic loading

Bases of Seismic Design Philosophy

Most common types of loads which a structure can experience throughout its life span are dead, live, snow etc., loads. These are static loads and their maximum values can easily be estimated using probabilistic approach. Earthquake loads are dynamic in nature and their maximum values cannot be estimated or calculated through this approach. The dynamic ground motions and dynamic properties of a structure determine the amplification or de-amplification of forces in terms of structural response. Resulting structural resistance can be interpreted in terms of energy dissipation capacity instead of strength capacity. The energy dissipation can be estimated in terms of product of forces and displacements. Hence, a structure's resistance against earthquake can be improved by either increasing its strength or increasing its deformability or by increasing both.

Several limit states criteria are used while designing a structure for static loads. After estimating maximum possible values for static loads, structure's response is analysed. Resistance capacity of individual members is calculated and this gives rules for determination of size of members so that external forces do not exceed the internal stresses of the members. Design criteria for earthquake loads is not that simple. The first step is to select a suitable maximum value for the seismic load. This maximum value mainly depends on the nature and seismic history of the area in which structure is being designed. To design a structure against maximum value of the seismic load is uneconomical as maximum value for seismic loads is rarely reached. Hence, a value smaller than the largest is considered at a certain level. Secondly, as the determination of exact values for seismic loads is almost impossible, individual members are designed and arranged in such a way that local damages can be avoided. Thus, maintaining the overall stability and integrity of the structure. This is achieved by making sure that member and connections have sufficient ductility so that they can withstand the dynamic seismic loading without collapse or damage (V Gioncu & Mazzolani, 2002).

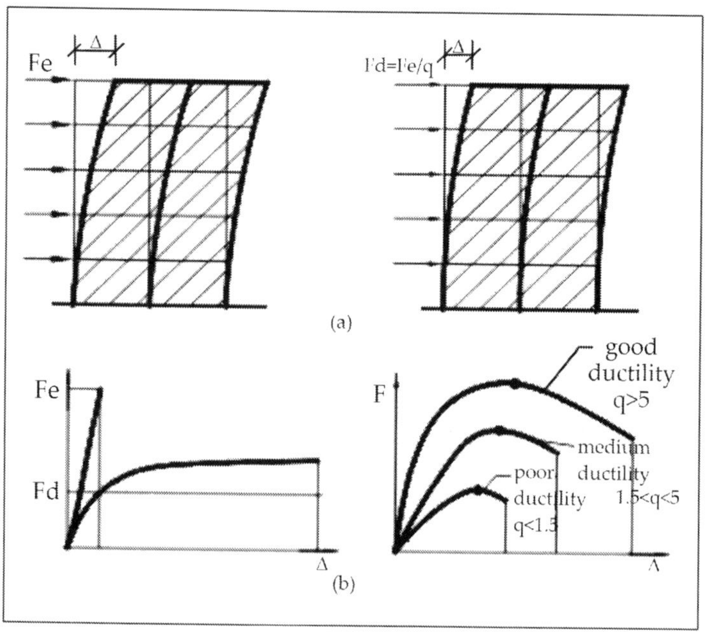

Figure 2.2: Illustration showing Structural stiffness

Criteria discussed above are considered as the foundation of modern design philosophy. According to this, the design horizontal force is obtained by dividing the maximum probable value for the seismic load that can be experienced by the structure by a reduction factor *q* Figure A. Values of q are given in seismic design codes. This factor is used so that plastic range of deformations for a structure can be increased to withstand stably against probable structural damage. This is defined as ductility and is shown in figure b.

Even after all above considerations, the nature of seismic actions and their explanation is still uncertain. Statistical analysis cannot be done with utmost certainty because records available for previous earthquakes are for too short time period compared to the geological time period. Still, effort can be made in increasing the resistive capacity of structures against seismic loading. All these considerations should be kept in mind during the construction phase as poor quality of construction might result in collapse of the structure no matter how safely it was designed on paper (ESDP, 2005).

Tasks of Earthquake Engineering

Now the basic considerations for seismic design have been explained, tasks of earthquake engineering can be clearly discussed. Earthquake engineering is considered as a new branch of Structural engineering because of its main scope of analysis of structure design and construction techniques against seismic loads. Different tasks which comprise the scope of earthquake engineers are discussed in this section.

While doing analysis for seismic design, it is very important to consider the type of earthquake. Following issues need to be addressed in this regard.

- Separate methodologies should be developed for inter and intra-plate earthquakes. This is due to the fact that ground motions can be different for both of these earthquake types. These methodologies can be developed using different ground motion records.
- Differences in pulse characteristics for near and far-field earthquakes need to be identified. Number of important cycles per second is very important for near-field earthquakes and

influence of local soil conditions should be investigated with regard to the stratification and dynamic properties of soil strata.

After establishing differences among different earthquake types, structural response against these earthquakes must be determined. This goal can be achieved by considering following parameters.

- Interaction between soil, foundation, structural and non-structural elements should be explained by considering actual properties of these elements individually and then modifying them with regard to earthquake level.
- Both horizontal and vertical components of seismic loads along with accidental torsions should be considered during the process of analysing the structure's response.
- Effect of neighbouring structures in case when seismic separation is not significant.

Following requirements need to be met in order to develop design methodologies explicitly for seismic loading.

- Equivalent lateral force analysis needs to be improved. This can be done by proposing different responses as a function of ground motions. Different response spectra for different types of earthquake are required.
- Proper distribution of loading should be determined to improve the static elastic-plastic analysis. Influence of different vibration modes also needs to be taken into account.
- Advancements are required for time-history analysis considering that sever earthquakes have longer period than moderate ones.

ISSUES FOR SEISMIC DESIGN

Motive behind earthquake architecture is the need to create such architectural designs which are supposed to address issues which can arise due to earthquakes. In other words, earthquake architectural designs are a mixture of earthquake engineering and architectural designs. Design of structure is expected to be aesthetically pleasing and most importantly, able to withstand all the resulting forces due to earthquakes and other types of imposed loadings. Some examples in which the structural conception gave important contribution to the architectural aesthetic are presented in Figure 1.8. In seismic design a key solution is to use bracing systems. If the architects have the ability to use these façade bracings in a creative modality, some very successful architectural expressions can be obtained.

The new aseismic systems used for energy dissipation have also an important influence on architectural configurations. Figure 1.9 shows some buildings where new bracing systems equipped with dissipative devices are located in the façades, enriching the architectural expression.

A very clear domain is where the bracing systems, applied to the inexpressive existing façades of buildings, make a real contribution to the improvement of the architectural value through the seismic strengthening of existing buildings (Charleson & Taylor, 2004; Victor Gioncu & Mazzolani, 2010).

SOME ETHICAL DILEMMA IN SEISMIC DESIGN

The Interests of economy, safety function and aesthetics rarely, if never, work together. More typically they run in opposite directions. In many situations the owners, structural engineers and architects play a different role with different opinions in relation to the problem of reducing the construction costs. This conflict of interest must be regulated by the *professional conscience.* For protecting the designer against this kind of pressures some professional codes have been elaborated by the State administrations. Structural engineers and architects, in their roles as building design professionals, enter into a special agreement with the general public: The State will protect their monopoly on certain segments the construction industry by allowing only registered engineers and architects to design buildings; the design professionals will have, in return, the duty to protect the public welfare in the built environment, by respecting the code rules. Therefore, the main purpose of a seismic code, as any other building code, is to protect life, health and public welfare. In ethical terms, such a code regulates the given actions on the basis of their consequences for the public(Spector, 1997).

The code mission is quite clear. The application, however, requires judgments and compromises due to the fact that a big gap still exists between the progressive knowledge and codification. Academics and researchers are devoted to research activity, but structural engineers are professionals and not researchers. So, they are basically conservative and resist the new concepts. A new knowledge is very difficult to implement in codes. Therefore, the code provisions must be, always, a compromise between new and old knowledge and procedures; otherwise the new methodology will be rejected by the designers.

In this situation, a progressive designer, who knows more about the actual structural behaviour during an earthquake than the codified one, has to face an ethical problem, whether he has to consider or not in practice his advanced knowledge. He has three choices:

- To accept the code provisions without any reserve, ignoring his knowledge;
- To reject the code of practice as inadequate, by substituting it with some personal provisions;
- To try to interpret the code provisions in the light of the new knowledge.

This attitude towards the use of the code is an ethical dilemma. The first position can be rejected because he renounces his ability as progressive designer. The second one is very dangerous because his personal judgment can lapse into arrogance. Therefore, the only way to solve this dilemma is the last attitude, to respect the spirit of the code, but introducing in design the new knowledge. The balance between old and new provisions is the dilemma, which must be solved by the progressive designers. Unfortunately, ethical theory offers little help in this dilemma (Spector, 1997).

INTERNATIONAL ACTIVITY

Seismology and Engineering Seismology

Seismology and Engineering Seismology are, in the truest sense, the global sciences, not limited by political boundaries. Thus, they can only be effectively practiced through international cooperation. When it was evident that seismic signals propagate through the Earth over large distances, the international organization for coordinating, collecting and distributing the data as well as promoting earthquake research became a pressing need. During two "International Conferences on Semiology" held in 1901 and 1903 in Strasbourg, the establishment of an "International Association of Semiology" (IAS) was proposed in 1904. This can be considered as the root of what finally became the *International Association of seismology and Physics of the Earth's Interior* (IASPEI) founded in 1951 (Adams, 2002; Henger, Berckhemer, & Seidl, 2002).

IASPEI promotes the study of earthquakes and other seismic sources, the propagation of seismic waves, and the Earth's internal structure, properties and processes. Scientists participating to IASPEI initiated to coordinate research and scientific exchanges which requires the cooperation in the earthquake prone countries. IASPEI scientific activities are categorized by the following themes: Earth structure and geodynamics earthquake sources – prediction and modeling, tectono-physics, earthquake hazard, risk and strong ground motions, seismological observation, interpretation of data, and education. Work on specific topics is carried out through commissions, sub-commissions, committees and working groups formed to meet specific needs of new, exciting problems as they emerge. One of these commissions is the *European Seismological Commission* (ECS), having the aim to promote seismological studies and projects in Europe, countries bordering the Mediterranean Sea and immediate neighbours: the area from the Mid-Atlantic Ridge to the Urals and the Arctic Ocean to Northern Africa. The activity of ECS is mainly focused in organizing every two years the General Assembly. The following objectives are of particular interest: to facilitate exchange of ideas and to organize personal meetings among scientists, to encourage the cooperation between individuals and organizations as well as between other European and non-European scientific and engineering communities by establishing Working Groups on seismological problems, to promote interdisciplinary studies involving seismology, and to organize training courses for young scientists.

Engineering Seismology scientific activity is framed in the general activity of IASPEI, focusing the seismological research topics applicable to the engineering communities as probabilistic seismic hazard assessment, their implications versus building codes for a seismic resistant design, and the earthquake risk assessment as a basis for risk management. Furthermore, engineering seismology basic research is performed to supply suitable datasets such as regional earthquake catalogues, numerical models and investigations on the earthquakes generating potential geological fault zones, and attenuation relations for potentially damaging ground motion parameters(Bolt, 2004).

Earthquake Engineering

Earthquake Engineering is a 20th Century development. In 1963 the *International Association for Earthquake Engineering* (IAEE) was founded, aiming at organizing World Conferences on Earthquake Engineering (WCEE). Since the IAEE foundation, these Conferences have been held every four years. The last ones, presenting very important progresses in seismic concepts were held in Madrid in 1992, Acapulco in 1996, Auckland in 2000, Vancouver in 2004, and Beijing 2008. These World Conferences represent a wide arena giving the opportunity to scientists, engineers, industrial professionals and government officials to present their scientific and engineering works, to exchange ideas and knowledge for the mitigation of seismic risk. They also provide a common platform for delegates for all over the world to initiate new cooperation. For instance, about 2500 delegates from about 73 countries were present during the 13th Vancouver Conference and 2300 technical papers were included in the Conference Proceedings (available as a CD-ROM containing approximately 27,000 pages).

In 1964 the *European Association for Earthquake Engineering* (EASS) (initially called European Commission for Earthquake Engineering) was founded during the Skopje Conference, having the same tasks as the IAEE organization. EASS organizes European Conferences on Earthquake Engineering (ECEE) every four years (in alteration with the WCEE Conferences). The last important ones were in Vienna in 1994, in Paris in 1998, in London in 2002, in Geneva in 2006, where for the first time a joint Conference of ECEE and ESC (European Seismological Commission) take place as the First European Conference on Earthquake Engineering and Seismology (ECEES). This was an excellent forum for idea exchanges between seismologists and earthquake engineers.

Similar chain of conferences is organized by *US Earthquake Engineering Research Institute* (EERI) as National Conference on

Earthquake Engineering (NCEE), the last ones being organized in Seattle in 1998, in Boston in 2002 and San Francisco in 2006 (to commemorate 100 years from the 1906 San Francisco earthquake). The EERI, founded in 1949 in Oakland (California), is a national nonprofit technical society of engineers, geoscientists, architects, planners, public officials and social scientists, having the main objective to reduce earthquake risk by advancing both the science and practice of Earthquake Engineering.

CONTROLLED SEISMIC SOURCES

Seismic waves created by blasts or vibrating controlled sources are one of the essential techniques for underground investigation in geophysics (notwithstanding a wide range of electromagnetic strategies, for example, instigated polarization and magnetotellurics). Controlled-source seismology has been utilized to guide salt arches, anticlines and other geologic traps in oil bearing rocks, flaws, shake sorts, and since quite a while ago covered goliath meteor holes. For instance, the Chicxulub Crater, which was brought on by an effect that has been ensnared in the termination of the dinosaurs, was restricted to Central America by investigating ejecta in the Cretaceous–Paleogene limit, and afterward physically demonstrated to exist utilizing seismic maps from oil investigation.

DETECTION OF SEISMIC WAVES

Seismometers are sensors that recognize and record the movement of the Earth emerging from versatile waves. Seismometers might be conveyed at the Earth's surface, in shallow vaults, in boreholes, or submerged. A total instrument bundle that records seismic signs is known as a seismograph. Systems of seismographs persistently record ground movements around the globe to encourage the observing and examination of worldwide quakes and different wellsprings of seismic action. Quick area of tremors makes tidal wave notices conceivable in light of the fact that seismic waves travel impressively quicker than tidal wave waves. Seismometers additionally record signals from non-tremor sources running from blasts (atomic and substance), to nearby clamor from wind or anthropogenic exercises, to perpetual signs produced at the sea depths and coasts incited by sea waves (the worldwide microseism), to cryospheric occasions related with extensive chunks of ice and icy masses. Above-sea meteor hits with energies as high as 4.2 × 1013 J (identical to that discharged by a blast of ten kilotons of TNT) have been recorded by seismographs, as have various modern mishaps and fear based oppressor bombs and occasions (a field of study alluded to as scientific seismology). A noteworthy long-haul inspiration for the worldwide seismographic checking has been for the location and investigation of atomic testing.

SEISM TECTONICS

Seism tectonics is the investigation of the connection between the quakes, dynamic tectonics and individual shortcomings of a district. It tries to comprehend which issues are in charge of seismic movement in a region by investigating a blend of local tectonics, late instrumentally recorded occasions, records of verifiable quakes and geomorphological proof. This data can then be utilized to measure the seismic danger of a range.

Methodology

A seism tectonic analysis of an area often involves the integration of disparate datasets.

Regional tectonics

A comprehension of the local tectonics of a zone is probably going to be gotten from distributed geographical maps, examine productions on the land structure and seismic reflection profiles, where accessible, expanded by other geophysical information.

So as to comprehend the seismic risk of a zone it is fundamental not exclusively to know where conceivably dynamic deficiencies are, additionally the introduction of the anxiety field. This is ordinarily gotten from a mix of quake information, borehole breakout investigation, coordinate anxiety estimation and the examination of topographically youthful blame systems. The World Stress Map Project gives a helpful online assemblage of such information.

Earthquakes

Instrumentally recorded occasions Since the mid twentieth century, adequate data has been accessible from seismometers to permit the area, profundity and extent of tremors to be figured. As far as recognizing the blame in charge of a seismic tremor where there is no evident surface follow, recording the areas of consequential convulsions for the most part gives a solid sign of the strike of the blame.

Over the most recent 30 years, it has been conceivable to routinely compute central components from teleseismic information. Lists of occasions with computed central components are presently accessible on the web, for example, the searchable index from the NEIC. As central systems give two potential dynamic blame plane introductions, other proof is required to decipher the starting point of an individual occasion. Albeit accessible for a confined day and age, in zones of direct to extraordinary seismicity there is most likely adequate information to describe the sort of seismicity in a region, if not all the dynamic structures.

Historical records

Endeavors to comprehend the seismicity of a region require data from tremors before the time of instrumental recording. This requires a watchful appraisal of recorded information regarding their dependability. By and large, all that can be determined is a gauge of the area and size of the occasion. Notwithstanding, such information is expected to fill the crevices in the instrumental record, especially in territories with either moderately low seismicity or where the rehash time frames for real quakes is more than a hundred years.

Field investigations

Information on the timing and magnitude of seismic events that occurred before instrumental recording can be obtained from excavations across faults that are thought to be seismically active and by studying recent sedimentary sequences for evidence of seismic activity such as seismites or tsunami deposits

Geomorphology

Seismically active faults and related fault generated folds have a direct effect on the geomorphology of a region. This may allow the direct identification of active structures not previously known. In some cases, such observations can be used quantitatively to constrain the repeat period of major earthquakes, such as the raised beaches of Turakirae Head recording the history of coseismic uplift of the Rimutaka Range due to displacement on the Wairarapa Fault in North Island, New Zealand.

SEISMIC PERFORMANCE ANALYSIS

Seismic investigation is a subset of auxiliary examination and is the figuring of the reaction of a building (or non-building) structure to quakes. It is a piece of the procedure of auxiliary outline, seismic tremor building or basic evaluation and retrofit (see basic designing) in areas where quakes are predominant.

As found in the figure, a building can possibly "wave" forward and backward amid a seismic tremor (or even an extreme wind storm). This is known as the 'central mode', and is the most reduced recurrence of building reaction. Most structures, nonetheless, have higher methods of reaction, which are particularly enacted amid seismic tremors. The figure just demonstrates the second mode, yet there are higher "shimmy" (strange vibration) modes. All things considered, the first and second modes tend to bring about the most harm much of the time.

The soonest arrangements for seismic resistance were the prerequisite to plan for a horizontal constrain equivalent to an extent of the building weight (connected at each floor level). This approach was embraced in the addendum of the 1927 Uniform Building Code (UBC), which was utilized on the west bank of the United States. It later turned out to be evident that the dynamic properties of the structure influenced the heaps produced amid a quake. In the Los Angeles County Building Code of 1943 an arrangement to change the heap in view of the quantity of floor levels was received (in light of research completed at Caltech in a joint effort with Stanford University and the U.S. Drift and Geodetic Survey, which begun in 1937). The idea of "reaction spectra" was produced in the 1930s, however it wasn't until 1952 that a joint board of trustees of the San Francisco Section of the ASCE and the Structural Engineers Association of Northern California (SEAONC) proposed utilizing the building time frame (the converse of the recurrence) to decide sidelong strengths.

The University of California, Berkeley was an early base for PC based seismic examination of structures, driven by Professor Ray Clough (who authored the term limited component. Understudies included Ed Wilson, who went ahead to compose the program SAP in 1970, an early "Limited Element Analysis" program.

Tremor building has built up a considerable measure since the good 'old days, and a portion of the more perplexing outlines now utilize uncommon seismic tremor defensive components either just in the establishment (base segregation) or appropriated all through the structure. Examining these sorts of structures requires particular unequivocal limited component PC code, which separates time into little cuts and models the real material science, much like normal computer games frequently have "material science motors". Large and complex structures can be demonstrated along these lines, (for example, the Osaka International Convention Center).

Equivalent static analysis

This approach characterizes a progression of strengths following up on a working to speak to the impact of quake ground movement, normally characterized by a seismic plan reaction range. It expects that the building reacts in its basic mode. For this to be valid, the building must be low-ascent and must not wind fundamentally when the ground moves. The reaction is perused from an outline reaction range, given the common recurrence of the building (either computed or characterized by the construction regulation). The appropriateness of this strategy is reached out in many construction standards by applying variables to represent higher structures with some higher modes, and for low levels of turning. To represent impacts because of "yielding" of the structure, many codes apply change figures that decrease the outline strengths (e.g. drive lessening elements).

Response spectrum analysis

This approach allows the numerous methods of reaction of a working to be considered (in the recurrence space). This is required in many construction laws for all with the exception of exceptionally basic or extremely complex structures. The reaction of a structure can be characterized as a blend of numerous extraordinary shapes (modes) that in a vibrating string compare to the "sounds". PC investigation can be utilized to decide these modes for a structure. For every mode, a reaction is perused from the plan range, in light of the modular recurrence and the modular mass, and they are then consolidated to give a gauge of the aggregate reaction of the structure. In this we need to ascertain the size of strengths every which way i.e. X, Y and Z and after that see the impacts on the building. Combination methods include the following:

- absolute, peak values are added together
- square root of the sum of the squares (SRSS)
- complete quadratic combination (CQC), a method that is an improvement on SRSS for closely spaced modes

The result of a response spectrum analysis using the response spectrum from a ground motion is typically different from that which would be calculated directly from a linear dynamic analysis using that ground motion directly, since phase information is lost in the process of generating the response spectrum. In cases where structures are either too irregular, too tall or of significance to a community in disaster response, the response spectrum approach is no longer appropriate, and more complex analysis is often required, such as non-linear static analysis or dynamic analysis.

Nonlinear static analysis

By and large, direct methodology is relevant when the structure is required to remain about versatile for the level of ground movement or when the plan brings about almost uniform appropriation of nonlinear reaction all through the structure. As the execution target of the structure infers more prominent inelastic requests, the instability with direct techniques increments to a point that requires an abnormal state of conservatism sought after presumptions and adequacy criteria to maintain a strategic distance from unintended execution. In this manner, techniques fusing inelastic examination can lessen the instability and conservatism.

This approach is otherwise called "weakling" examination. An example of powers is connected to an auxiliary model that incorporates non-straight properties, (for example, steel yield), and the aggregate compel is plotted against a reference uprooting to characterize a limit bend. This can then be joined with a request bend (regularly as a quickening relocation reaction range (ADRS)). This basically decreases the issue to a solitary level of flexibility (SDOF) framework.

Nonlinear static systems utilize proportional SDOF basic models and speak to seismic ground movement with reaction spectra. Story floats and part activities are connected therefore to the worldwide request parameter by the weakling or limit bends that are the premise of the non-direct static methods.

Nonlinear dynamic analysis

Nonlinear dynamic investigation uses the blend of ground movement records with an itemized auxiliary model, subsequently is fit for creating comes about with moderately low instability. In nonlinear dynamic examinations, the definite auxiliary model subjected to a ground-movement record produces assessments of segment mishaps for every level of opportunity in the model and the modular reactions are consolidated utilizing plans, for example, the square-root-entirety of-squares.

In non-straight unique examination, the non-direct properties of the structure are considered as a major aspect of a period area investigation. This approach is the most thorough, and is required by some construction regulations for structures of bizarre design or of exceptional significance. In any case, the ascertained reaction can be extremely delicate to the qualities of the individual ground movement utilized as seismic information; along these lines, a few examinations are required utilizing distinctive ground movement records to accomplish a solid estimation of the probabilistic conveyance of auxiliary reaction. Since the properties of the seismic reaction rely on upon the power, or seriousness, of the seismic shaking, a far-reaching appraisal requires various nonlinear dynamic examinations at different levels of force to speak to various conceivable quake situations. This has prompted the rise of techniques like the Incremental Dynamic Analysis.

3
CHAPTER

ROLE OF GEOLOGY IN EARTHQUAKE STUDIES

INTRODUCTION

The commitment of geography to seismic danger relief is that the past is the way to what's to come. Where a seismic tremor has occurred previously, it is probably going to happen once more. The essential part of the geologist is to find paleo-tremors utilizing paleo-seismology, find and portray the deficiencies that create them and build up a quantitative comprehension of the style, extent, recurrence and recentness of quakes on those issues as a manual for their future conduct. Topography, along these lines, is both locative and prescient. It can be utilized to gauge the extent of the seismic tremor, the area of the blame burst, the surface twisting foreseen where the blame does not achieve the surface, and the nearby site enhancement considers that will affect solid ground shaking and soil disappointments. Topography gives the quantitative establishment on which different parts of seismic peril alleviation are based.

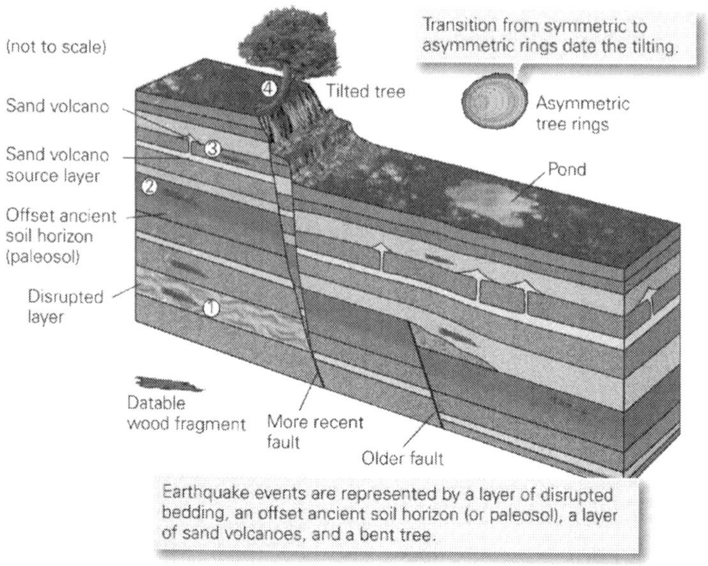

Figure 3.1: Illustration of Relation of Geology and Earthquake

Early investigations of seismic tremor dangers were directed after significant quakes in Marlborough, New Zealand, in 1888; Mino-Owari, Japan, in 1891; Assam, India, in 1897 and San Francisco, California, in 1906. In these examinations, geologists assumed driving parts. The cautious blame crack portrayals of Alexander McKay of New Zealand, Bunjiro Koto of Japan and G.K. Gilbert of the United States set a standard that has once in a while been surpassed (see individual vignettes in Yeats et al., 1997). These reviews concentrated at first glance impacts of the seismic tremor, including crustal distortion. Beside Gilbert, however, these scientists did not consider the suggestions in future arranging and building outline.

In consequent decades, geologists kept on assuming a part, yet a subordinate one because of two advancements. One was the Wood-Anderson seismograph, planned in the 1920s, which brought about an undeniably imperative pretended by geophysicists. Another was the appropriation in California of the principal nearby construction laws mulling over tremor resistance, beginning after the 1925 Santa Barbara quake and quickening after the Long Beach seismic tremor of 1933. People in general objection after the Long Beach quake prompted the reception of the Field and Riley Acts by the State of California, bringing about an inexorably imperative part for auxiliary architects. For the half century that took after the 1906 San Francisco seismic tremor, geologists dealing with quake related issues were completely college teachers or representatives of the U.S. Topographical Survey (see recorded rundowns by(Yeats & Gath, 2004)).

The improvement of building regulations managing the nature of building development was not joined by statutes managing the security of building destinations until 1952, when the City of Los Angeles received the principal evaluating mandate in the United States and set up a reviewing segment inside the Department of Building and Safety. This move was activated not by seismic tremors but rather via avalanches, which were an expanding issue as the growing populace of Los Angeles moved into the encompassing foothills. By 1963, improvement ventures were required to be bolstered by both geotechnical building and designing topography reports, and evaluating operations must be administered by both a dirt build and a designing geologist. This prompted the foundation of topographical counseling firms and to the work of geologists in bigger counseling designing firms. For instance, F. Beach Leighton, an educator of topography at Whittier College, set up a geographical counseling hone in an extra room of his home, then extended to a huge association with workplaces all through southern California.

California now requires geographical counsel chiefly because of two demonstrations of the lawmaking body: the Alquist-Priolo Earthquake Fault Zoning Act, first embraced in 1972 to control development on or close dynamic deficiencies and the Seismic Hazard Mapping Act of 1990, managing land subject to liquefaction and tremor initiated arrive sliding. The state office given important obligation regarding these laws is the California Geological Survey (some time ago the Division of Mines and Geology), which, what's more, helps neighborhood governments in embracing Seismic Safety Elements as a feature of their General Plans, in view of a law gone in 1971.

California has the most developed seismic tremor defensive enactment in the United States, maybe on the planet (Yeats, 2001), however different states have not taken after its lead in assessing the security of building locales. Oregon, for instance, has overhauled its construction regulations near California gauges, however its property utilizes laws ensuring against geologic perils are basically unaltered. Subsequently, building locales in Oregon, similar to those in different states in mischief's way from seismic tremors, succumb to what has turned out to be known as Slosson's Law, progressed by topographical advisor and previous California State Geologist James Slosson: The nature of expert work will sink to the most reduced level that legislature will acknowledge.

This section presents cases from counseling rehearse in which geography was vital, even basic, to seismic danger moderation. Most illustrations are from California, where topographical reviews are required, however the section is composed with the target that a cutting-edge investigation anyplace on the planet will incorporate geographical components, regardless of whether it is required by state or nearby law. For more insights about the land foundation, allude to McCalpin (1996) and Yeats et al. (1997).

PALEO-SEISMOLOGY

Customers need to know when the following seismic tremor is probably going to strike close to the venture site, and how much harm the quake will do. Here and now expectation is not yet not too far off and may never be in our lifetimes. So, we should depend on probabilistic conjectures that give the probability of a quake striking in a time allotment of intrigue, normally the life of the venture being assessed, or deterministic gauges that give the biggest tremor (measured in pinnacle ground increasing speeds and speeds and reaction spectra) that is probably going to influence the venture zone. Deterministic figures are valuable for basic offices, for example, substantial dams or atomic power plants.

It is standard practice to report the instrumental and authentic seismicity of the district encompassing a venture site as a sign of potential quake danger. However, this practice gives too short a period test to be of noteworthy esteem. Seismographs have been set up for just a century, and present day broadband seismographs have been set up for just a couple of decades. The authentic record for, say, California, is dependable for just around 150 years aside from a couple populace focuses along the drift, where it is somewhat more. This day and age is too short to distinguish the latest seismic tremor on the Cascadia Subduction Zone, the southern San Andreas blame or quakes of M>7 in the Los Angeles metropolitan locale (Dolan & Rockwell, 2001). Moreover, instrumental seismicity, notwithstanding utilizing best in class seismographs, is a poor manual for potential danger with the exception of in the broadest terms. A large portion of the 1857 crack zone and some portion of the 1906 break zone on the San Andreas Fault have low seismicity (Yeats, 2001). The Cascadia subduction zone off the bank of the Pacific Northwest has the potential for a tremor of Mw 9, however not very many quakes have been recorded on the subduction zone, and basically none north of California. The reason is that seismic tremors, huge and little, mirror the sudden arrival of strain, though measuring the development of strain is more imperative in tending to worries about potential quakes. In the event that the blame zone being referred to is solid, there is no motivation to accept that a tremor on that blame will transmit its punch by little quakes, albeit a few seismic tremors on issues in China and Greece were gone before by enough foreshocks to caution the nearby populace (Yeats, 2001).

Both recorded and instrumental seismicity are helpful in describing shortcomings that have short quake return times, similar to the San Andreas Fault at Park field where the repeat interim is a couple of decades. Authentic seismicity is valuable for shortcomings for which the repeat interim is essentially shorter than the verifiable record, for example, the Motagua blame in Guatemala, the North Anatolian blame in Turkey and a few blames in Iran and China. Be that as it may, for most blames, particularly those in the western United States, neither recorded nor instrumental seismicity can give data about a total tremor cycle. The geologic record, which is thousands or even a huge number of years long, can traverse an entire cycle and may traverse a few cycles. The geologist's obligation, then, is to decide the slip rate and the seismic tremor repeat interim of flaws close to the site (line wellsprings of Reiter [1990]). Moreover, it might be important to decide the repeat interim for a district containing a few shortcomings (territory source), and take into consideration blames that have not yet been distinguished (drifting quake). The slip rate can be controlled by the balanced of components of known age, with the age decided by and large by radiocarbon dating or different methods, for example, optically fortified glow, which does not require carbon. The component might be a balanced stream channel over the San

Andreas Fault, a counterbalance marine stage and shoreline precipice or a balance skyline in a bulldozer or excavator trench exhuming over a blame. Slip rates decided thusly can be contrasted and rates gotten from plate tectonics (which depend on time allotments of many thousands to a huge number of years) and with strain amassing rates (which depend on rehashed reviews utilizing the Global Position System, crossing a time span of quite a long while). For metropolitan Los Angeles, slip rates have been resolved on issues over the long haul (a large number of years) to the fleeting (a huge number of years to a couple of years). For a few blames, the Southern California Earthquake Center has confirmed that the fleeting rate, which is basic in danger estimation, varies essentially from the long-haul rate.

Urbanization has anticipated subsurface unearthings crosswise over many blames in Los Angeles, albeit subsurface examination remains an attractive objective. In the San Francisco Bay Area, the BAPEX (Bay Area Paleoseismic Experiment) venture is an engaged push to refine the comprehension of here and now slip rates and repeat interims. In view of paleoseismic confirmation assembled to date, the Bay Area is under a long-haul conjecture of two possibilities out of three for a seismic tremor of greatness more noteworthy than 6.7 in the following 30 years, and one shot out of three for a quake of this size on the intensely urbanized northern Hayward blame and the nearby Rodgers Creek blame (Working Group on California Earthquake Probabilities, 1999).

Singular seismic tremors in a trench uncovering can be perceived by wedges of colluvial trash shed from a blame scarp amid a quake (Figure 3.2b), liquefaction elements, for example, sand embankments (Figure 3.2c) and marine shorelines all of a sudden inspired in the previous couple of thousand years amid a seismic tremor. A blame strand uncovered in a trench can end upward against a more youthful unfaulted sedimentary layer (Figure 3.2c), or layers tilted amid a tremor can be overlain by level lying dregs. The age of the quake is sectioned by dating the most youthful blamed or tilted dregs (layer B, Figure 3.2c) and the most established unfaulted or level lying residue (layer A, Figure 3.2c). Utilizing these criteria in trench exposures over the San Andreas blame south of Palmdale at Pallett Creek and adjacent regions, eight ancient seismic tremors reaching out back 1300 years were distinguished notwithstanding the crack amid the 1857 Fort Tejon quake of M 7.9

(Sieh, 1978; Sieh and LeVay, 1998).

The earthquake recurrence time t is equal to the slip per earthquake **s** divided by the slip rate on the fault, **v**:

T = s/v

Slip per event is determined by field observations after a surface-rupturing earthquake, by observations of displacement in a trench or series of trenches and by offsets of small streams or cultural features across a fault.

In the event that tremor repeat were occasional, determining the following one would be clear. However, paleoseismic time histories demonstrate that seismic tremor repeat is not occasional. Also, at times, most outstandingly along the focal San Andreas blame, quakes are not trademark—that is, a similar portion of blame may create tremors of various blame removals, inferring diverse extents, at various circumstances (Grant, 1996,(Minor & Grant, 1996)).

The huge number of quakes on the Central Nevada Seismic Zone east of Reno (1903, 1915, 1932, 1934 and four in 1954) does not correspond with the long-haul tremor repeated in the interim on this present zone's flaws (which is a large number of years for a similar blame portion). This irregularity recommends that quakes group in time. Co-seismic strain discharge over the Eastern California Shear Zone over the most recent 60 years has been commonly more than that over the close-by San Andreas blame, however the slip rate over the Eastern California Shear Zone is just a little part of that over the San Andreas blame.

The extent of a paleo-seismic tremor can be evaluated on the premise of the length of the blame source and the most extreme removal amid an individual quake. Wells and Coppersmith (1994) arranged relapse connections among extent, crack length and relocation for 216 recorded seismic tremors. Despite the fact that there are contrasts between areas, these connections are utilized to gauge greatness from land information.

One of the quakes on the Eastern California Shear Zone, the Landers tremor of M 7.3 in 1992, gave prove that a seismic tremor can bounce from blame to blame. Trench unearthings crosswise over flaws in the Los Angeles Basin demonstrate confirmation of blame relocations that are too huge for the seismic tremor to have been restricted to the blame that was trenched (Rubin et al., 1998). The worry about crack on various issues is that if the seismic tremor is bigger, the region of solid shaking is bigger, and the length of solid shaking is longer. This prompts a seismic tremor break may course from blame to blame, similar to catches ripping off a shirt. To assess this plausibility, it is important to get subsurface data on all deficiencies in a locale, particularly the Los Angeles area where repeat interims on most blames are measured in a large number of years and accordingly subject to thorough paleo-seismic examination.

SURFACE RUPTURE

Seismic tremor shortcomings are of three sorts. Strike-slip deficiencies are portrayed by movement parallel to the Earth's surface, balancing streams) and social components. The best-known case is the San Andreas Fault, with surface crack in the Fort Tejon tremor of 1857 and the San Francisco quake of 1906. Typical issues are portrayed by the Earth's covering pulling separated, bringing on one side of the blame, the upper square, called the hanging-divider, to drop down regarding the opposite side. Cases incorporate surface burst in the Pleasant Valley tremor of 1915 and the Dixie Valley seismic tremor of 1954, both in western Nevada. Switch flaws are described by the Earth's hull being constrained together so that the hanging divider rides up over the other square. Cases incorporate surface burst going with the San Fernando (Sylmar) tremor of 1971 and the Chi-Chi, Taiwan, quake of 1999.

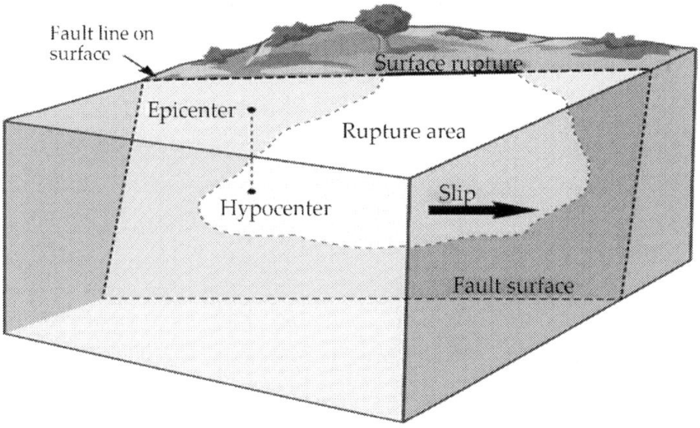

Figure 3.2: Surface rupture due to Earthquake

The Alquist-Priolo (A-P) Act emerged as a result of harm from the turnaround blame that cracked crosswise over private rural areas in the San Fernando Valley in the 1971 San Fernando (Sylmar) seismic tremor. The eastern segment of the surface burst was known to be a blame yet was not perceived as dynamic; the western area was not perceived as a blame before the seismic tremor. The Act was hastened to some degree by uncontrolled advancement over the San Andreas Fault in the San Francisco Bay Area and in suburbia of San Bernardino and over the Hayward and Calaveras blames east of San Francisco Bay. Most youthful blames in California have been distinguished and portrayed on the premise of the age of their latest crack (Jennings, 1994), and some of these shortcomings have been zoned under A-P. To fall under the ward of A-P, the blame must be shown as dynamic, in light of proof that the blame has experienced surface crack in the previous 11,000 years. (As Clarence Allen of Caltech watched, a blame that has moved as of late is probably going to move once more. Expressed an alternate way, what has occurred, can happen.) Notwithstanding being dynamic, a blame zoned under A-P must be obviously characterized in the field by a geologist (Hart and Bryant, 1997).

A-P zone limits are set at 500 ft. from most dynamic blames yet can be as low as 200 ft. for less huge deficiencies. Inside an A-P zone, exceptional reviews must exhibit to an elevated expectation that a venture inside the zone is sheltered against surface burst. By and by, the most widely recognized investigative system is to unearth a trench over a proposed fabricating site and guide the dividers of the trench for the nearness of conceivable dynamic deficiencies (Figure 3.2b-d). Dating the recentness of blame action requires the nearness of unfaulted datable silt. In the event that these residues are more seasoned than 11,000 years, no blame at the site would be zoned under A-P, regardless of whether uncovered by the trench or not. The law builds up a misfortune from a zone of surface blaming in view of suggestions by the geologist. By law, this mishap is 50 ft., however the misfortune can be diminished or expanded on the proposal of the venture geologist. How wide ought to the mishap be around a blame distinguished under A-P as dynamic? At most venture locales, the distributed blame area is not itemized enough to settle on development choices. The geologist, through watchful investigation of the geology and subsurface confirmation from trench unearthings and boreholes, must find and portray the site-particular peril of the blame zone, which may comprise of a system of discrete shortcomings, any of which could harm or wreck an adjacent building.

A-P has been reprimanded for assaulting the wrong issue. In the 1971 tremor, the harm from surface burst was significantly not as much as the harm from different causes, for example, solid shaking or liquefaction. The following three urban tremors in California—1987 Whittier Narrows, 1989 Loma Prieta and 1994 Northridge—were not joined by surface crack, yet misfortunes from each of the last two seismic tremors were measured in billions of dollars. In any case, the Chi-Chi, Taiwan, tremor of September 21, 1999 was joined by numerous miles of surface break on the Chelungpu switch blame in created territories. About entire obliteration come about along the blame crack with incredible death toll, especially on the up thrown (hanging wall) side of the blame (Figure 3.3). The blame had been mapped preceding the tremor; if A-P had been in actuality before improvement, many lives would have been spared and property misfortunes would have been incredibly lessened (Yeats, 2001). Numerous California urban communities and areas, through their open security components, have received a more stringent elucidation to incorporate a greater number of deficiencies than those zoned by the state under A-P.

BLIND THRUSTS

The Whittier Narrows, Loma Prieta and Northridge seismic tremors, together with a prior quake in 1983 that devastated the vast majority of the downtown segment of Coalinga, California, were not joined by surface break. These seismic tremors actuated switch flaws, yet blame crack halted a few miles underneath the surface. Such blames are called daze pushes. They are not very much characterized at the surface, so they are not zoned under A-P. The four tremors said above struck deficiencies that were not perceived at the time as dynamic; truth be told, the Northridge seismic tremor blame was not known by any means. However, these flaws for the most part have surface expression as sedimentary layers inspired and collapsed into anticlines (Figure 3.4). Portrayal of the level of movement of visually impaired pushes requires the utilization of oil industry well logs and seismic profiles and in addition topographic elements, for example, stream patios and marine shorelines that have been up twisted amid numerous past visually impaired push quakes (Stein and King, 1984; Anderson, 1990; Stein and Ekström, 1992; Shaw and Shearer, 1999). The Northridge dazzle push came up underneath a surface turn around blame dunking the other way. The confirmation of its nearness, found later, was the inspire of both pieces of the surface blame (Yeats and Huftile, 1995; Huftile and Yeats, 1996). Examination of the overlying fold allowed a gauge of the long-haul slip rate on the visually impaired push (Davis and Namson, 1994; Huftile and Yeats, 1996). On the off chance that the surface proof of the visually impaired push is restricted to wide distorting of the surface in anticlines, then the primary impacts (beside solid shaking) are the adjustment in slope of railroad lines and pipelines. Be that as it may, anticlinal distorting is generally joined by auxiliary shortcomings, coming about because of sedimentary layers slipping along bedding as the layers are collapsed (flexural-slip blaming).

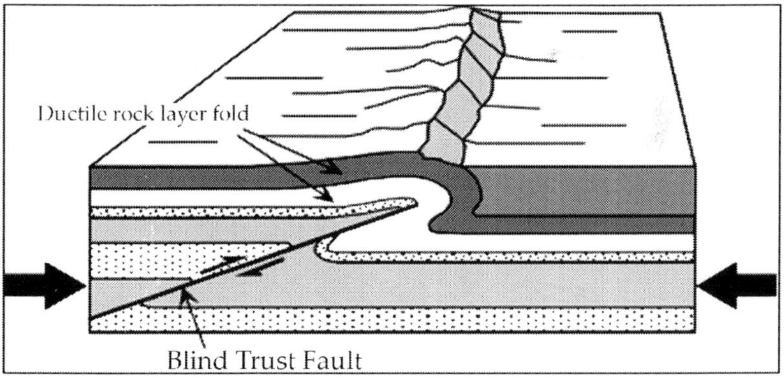

Figure 3.3: A blind thrust fault

Anticlinal warping also stems from the formation of faults or parasitic folds resulting from *bending moment* imposed during folding (Yeats et al., 1981, 1997; Oskin et al, 2000]. Treiman (1995) reported flexural-slip reverse faults that broke the ground surface of building pads graded for a housing development in Santa Clarita, California during the 1994 Northridge earthquake. The Coyote Pass escarpment in East Los Angeles is a steep, localized fold scarp that accompanied uplift of a large anticline to the north (Oskin et al., 2000).

STRONG GROUND MOTION

Delicate surficial residue has a tendency to open up seismic waves, bringing about the bowl of jello impact. This was shown by contrasting the shaking at Fort Mason, west of Fisherman's Wharf in San Francisco, with the shaking somewhere else in the city amid the 1906 quake. Chief M.L. Walker at Fort Mason felt the 1906 seismic tremor yet then backpedaled to rest, feeling that it was close to a mellow shaker. Seismograms of post-quake tremors of the 1989 Loma Prieta tremor recorded much lower speeds of seismic waves at Fort Mason than in the Marina District more remote west, a resound of 1906.

Stronghold Mason is alluded to as a stone site, and the Marina District as a dirt site. One of the most noticeably bad illustrations is Mexico City, based on delicate earth, sediment and sand of old Lake Texcoco. A tremor in 1985 on the Pacific Coast, many miles far from Mexico City created extraordinary harm in Mexico City, as a result of the expansive populace thought there and the intensification of seismic waves by the delicate residue of Lake Texcoco, Mexico City's bowl of jello.

Recognizable proof of low-lying regions subject to solid ground movement is encouraged by shear-wave speed estimations in boreholes, yet the geologist can characterize the breaking points of such territories by examination of the depositional setting of silt basic the site being referred to in view of topographic maps, aeronautical photographs, centers taken in boreholes and high determination, shallow-infiltration seismic profiles. Additionally, the design of bowls in view of wells and more profound infiltration seismic profiles acquired by the oil business is critical in light of the fact that sedimentary bowls trap quake waves and make them resound off bowl edges, expanding the length of solid shaking. The Los Angeles bowl is a case of a profound sedimentary bowl that could trap seismic waves (Olsen et al., 1995).

EARTHQUAKE-INDUCED LANDSLIDES

Vast tremors create a large number of avalanches. The Northridge tremor activated more than 11,000 avalanches, generally in the Santa Susana Mountains and in the slopes north of the Santa Clara River (Harp and

Jibson, 1996). Some of these are moderately little, yet a couple have measurements measured in miles and would be calamitous in the event that they struck an urban zone. The El Salvador tremor of January 13, 2001 (Mw 7.6) set off an avalanche in the Las Colinas center pay suburb of San Salvador, more than 60 mi far from the epicenter, executing around 400 individuals (Lomnitz & Elizarrarás, 2001). Some quake activated avalanches happen seaward and result in tidal waves. The immense 1964 Alaska tremor brought on submarine avalanches at Seward and Valdez, producing wave waves up to 160 ft. high that crushed the waterfronts of both towns (Hampton et al., 1993). Area and portrayal of tidal waves produced by seaward avalanches rely on upon nitty gritty examination of seaward geology, requiring multi bar side sweep sonar and bathymetric symbolism.

This can be exceptionally noteworthy in researching basic offices along the drift, for example, atomic power plants and condensed flammable gas plants. Avalanches, similar to dynamic flaws, have a tendency to happen in spots that have had avalanches previously, usually in ancient circumstances. These antiquated avalanches can be perceived by the geologist in view of flying photograph examination and field work. The starting point can be amazingly overwhelming precipitation rather than a tremor, however the outcome to manufactured structures is the same. The geologist can likewise perceive elements that prompt avalanches. For instance, cutting into a slant in which bedding plunges in an indistinguishable heading from the slant can trigger downslope development, a noteworthy issue along the focal Oregon drift at Newport and on the north side of the Santa Clara River close Ventura, California. Another component prompting avalanches is the nearness of feeble residue at the surface, for example, wind-blown sediment in the Portland Hills in Portland, Oregon, and cold silt in shoreline bluffs close Seattle, Washington. The Las Colinas slide in El Salvador occurred in fine wind-blown fiery remains stores that because of their low quality, have the potential for slant disappointment (Lomnitz & Elizarrarás, 2001).

Figure 3.4: Landscape view of the cliff landslide caused by the earthquake at Serempah village, central Aceh district

Scott Burns, an engineering geologist at Portland State University, uses the three-strike rule. Rule 1 is unstable soil, rule 2 is a steep slope and rule 3 is the landslide trigger, either earthquake shaking or heavy rainfall that saturates the ground. By careful selection of building sites, rules 1 and 2 can be avoided so that neither an earthquake nor heavy rainfall will trigger a landslide on a project site.

Liquefaction

Liquefaction occurs when the soil beneath the surface actually behaves like a fluid under strong shaking. The breakup of the Turnagain Arm housing development in a suburb of Anchorage in the 1964 Alaska earthquake was largely due to the liquefaction of a sand layer in the Bootlegger

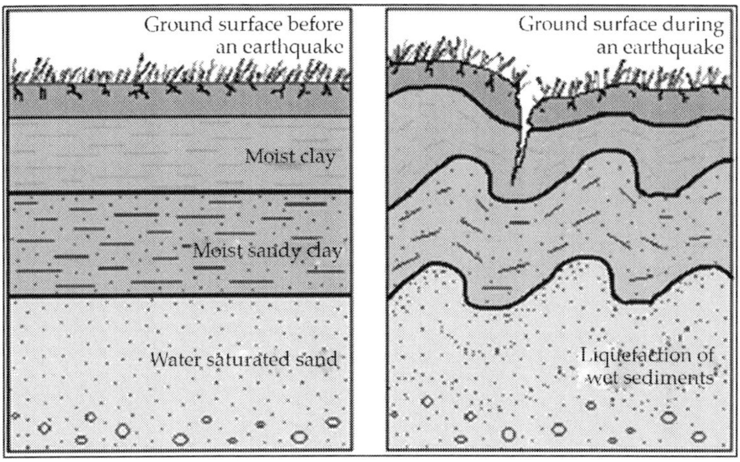

Figure 3.5: Soil Strata before and after an earthquake showing liquefaction

Mud fundamental the subdivision. A great part of the harm in the Marina District of San Francisco in the 1989 Loma Prieta tremor was the consequence of liquefaction of the manufactured fill emplaced after the 1906 seismic tremor. Dregs in the floodplain of the Los Angeles River and Ballona Creek melted in the 1994 Northridge seismic tremor, pulverizing structures and snapping sewer and water lines. The most marvelous surface impacts of the February 28, 2001 Nisqually, Washington seismic tremor (Figure 3.5) were brought on by liquefaction, incorporating harm to subdivisions in Olympia and to Boeing Field. In a few cases, liquefaction amid the Nisqually seismic tremor occurred at an indistinguishable spot from liquefaction because of quakes in 1949 and 1965. These liquefaction-inclined ranges had been recognized preceding the seismic tremor on maps distributed by the Washington Division of Geology and Earth Resources (Coe et al., 2004).

These dregs, which are subjected to liquefaction, are spotless, attachment-less, water-immersed sand by and large inside 30 ft. of the surface, kept in the Holocene Epoch, in the course of the most recent 10,000 years. This liquefiable silt might be overlain by a generally impermeable layer, an earth top. On the off chance that the dregs are found on inclines, even those as low as 1°, the impermeable layer, and structures based on it, can move as a sidelong spread, breaking underground utility lines. Softened underground gas lines up the Marina District touched off in 1989, creating an expansive fire. In spite of the fact that the powerlessness of a silt layer can be measured by the Standard Penetration Test (SPT), ranges liable to contain liquefiable dregs can be distinguished by the geologist utilizing investigation of air photographs, point by point digitized geography, normal exposures and deliberately chose boreholes. For instance, liquefaction impacts of the New Madrid, Missouri seismic tremors of 1811–1812 are as yet obvious on air photographs taken a century and a half later ((Yeats & Gath, 2004)).

4 CHAPTER

EFFECTS OF EARTHQUAKE

INTRODUCTION

The impacts of seismic tremors on the constructed condition can be studied under two classes: coordinate impacts and aberrant impacts. Solid ground shaking underneath a structure is the essential impact of a quake, which influences the structures in the territory hit by the seismic tremor. A blame burst crossing a long structure or a foundation arrange (life saver, for example, a viaduct, a pipeline framework or a street system is likewise considered as an immediate impact of a seismic tremor. Aberrant impacts are fundamentally geotechnical disappointments therefore of solid ground shaking, which thus influences the structures situated in that geotechnical condition. Liquefaction, horizontal spread and seismic tremor actuated avalanches are cases of geotechnical events. Tidal waves and fire due to earthquakes are likewise considered impacts of tremors. Backhanded impacts are in no way, shape or form less critical. Truth be told, the loss of life and financial misfortunes because of tidal waves amid the 2004 Indonesia (Sumatra) and 2011 Japan (Tohoku) seismic tremors were much more serious than the misfortunes because of the immediate impacts of these quakes.

Figure 4.1: A Dramatic picture of destructive tsunami of 2011 Japan earthquake

Direct Shaking Hazards and Human-Made Structures

Most seismic tremor related passings are brought about by the fall of structures, and the development rehearses assume a huge part in the loss of life of a quake. In southern Italy in 1909 more than 100,000 individuals died in a seismic tremor that struck the area. Half of the general population living in the locale of Messina were executed because of the effortlessly collapsible structures that commanded the towns of the district. A bigger tremor that struck San Francisco three years prior had killed fewer individuals (around 700) on the grounds that building development practices were diverse sort (transcendently wood). Survival rates in the San Francisco quake were around 98%, compared to the Messina seismic tremor which was in the vicinity of 33% and 45% (Raup, 1997). Building practices can have a significant effect in quakes, even a direct burst underneath a city with structures caught off guard, can create a huge number of losses.

Albeit likely the most critical, direct shaking impacts are by all account not the only peril related with seismic tremors, different impacts, for example, avalanches, liquefaction, and tidal waves have likewise had vital influence in obliteration created by quakes.

GEOLOGIC EFFECTS OF SHAKING

When we talked about tremor power we examined a portion of the essential elements that influence the sufficiency and term of shaking delivered by a quake (tremor measure, remove from blame, site and territorial topography, and so forth.) and as you know, the shaking created by seismic waves can bring about harm structures or make structures crumple. The level of harm done to a structure relies on upon the plenteousness and the span of shaking.

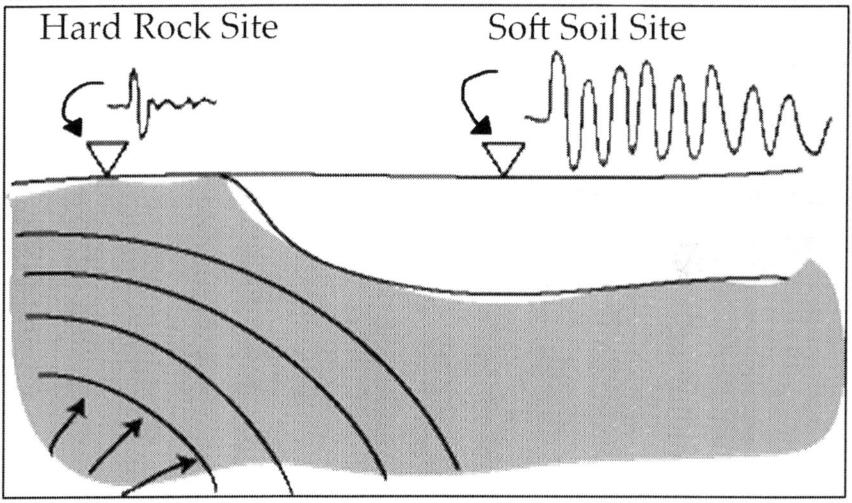

Figure 4.2: A graphical representation of Hard rock sliding under soft soil

The amplitudes are biggest near extensive seismic tremors and the term for the most part increments with the measure of the seismic tremor (bigger shudders shake longer since they burst bigger territories). Territorial geography can influence the level and term of shaking; however, more imperative are neighborhood site conditions. In spite of the fact that the procedure can be confounded for solid shaking, for the most part shaking in delicate residue is bigger and longer than when contrasted and the shaking experienced at a "hard shake" site.

Figure 4.3: A photo of highway in japan after 2011 Earthquake

Preparing Structures for Shaking

The initial phase to prepare structures for shaking is to see how structures react to ground movements; this is the field of study for tremor and basic architects. At the point when the ground shakes, structures react to the increasing velocities transmitted starting from the earliest stage the structure's establishment. The inactivity of the building (it needs to remain very still) can bring about shearing of the structure which can focus weights on the powerless dividers or joints in the structure bringing about disappointment or maybe add up to fall. The sort of shaking and the recurrence of shaking relies on upon the structure. Tall structures have a tendency to intensify the movements of longer period movements when contrasted and little structures. Each structure has a reverberation recurrence that is normal for the building. Foreseeing the exact conduct of structures is confused, a general guideline is that the time of reverberation is about equivalent to 0.1 times the quantity of stories in the structure. In this way Macelwane Hall reverberates at around 0.3 seconds time frame, and Griesedeck at around 1.4 seconds.

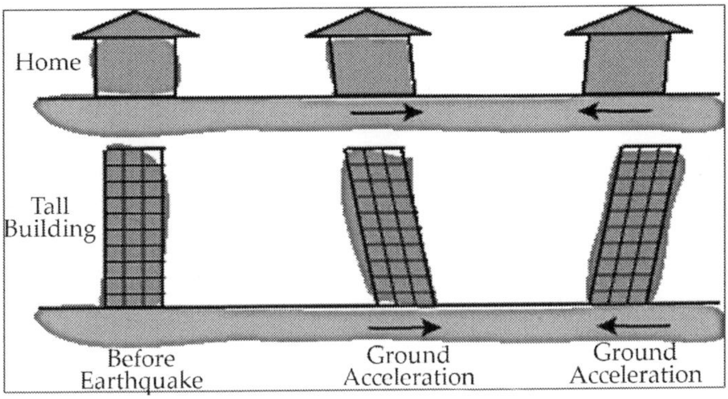

Figure 4.4: A Demonstration of vibration of building during Ground shaking of earthquake

Taller structures additionally tend to shake longer than short structures, which can make them moderately more powerless to harm. Luckily numerous tall structures are developed to withstand solid winds and a few insurances have been taken to decrease their propensity to shake. Also, they can be made impervious to quake vibrations. In numerous areas of constrained assets and additionally old structures, the structures are not suited to tremor incited strains and fall of adobe-style development has brought about a large number of passings in the most recent decade. The most exceedingly terrible conceivable structure for tremor areas is the unreinforced stone work (which is normal in the St. Louis region).

Engineering Characterization of Ground Motion

"The basic data of earthquake engineering are the recordings of ground accelerations during earthquakes. A knowledge of the ground motion is essential to an understanding of the earthquake behavior of structures" (Housner, 1970a). Recorded ground movement time arrangement contain important qualities and data that are utilized straightforwardly, or by implication, in seismic investigation and outline. Parameters, for example, crest ground movement values (speeding up, speed and dislodging), measures of the recurrence of the ground movement, the term of solid shaking and different force measures assume vital parts in seismic assessment of existing offices and outline of new frameworks.

ESTIMATING HAZARDS

Planning structures (either new or old) for seismic tremors is costly and the level of speculation is a social and political choice. The decision of building configuration is a tradeoff between appearances, work, structure, quality, and obviously, cost. Norms are founded through the foundation of Building Codes, which control the outline and development of structures. A large portion of our construction laws are intended to ensure first the building inhabitants, and second the building respectability. Construction regulations are generally drafted to meet the requests of the normal shaking in a given district that are compressed by seismologists and quake designs in risks maps. Hazard maps are constructed by examining

- The earthquake history of the region to estimate the probability of an earthquake
- The expected shaking intensity produced by the earthquake (often expressed as a peak acceleration)
- The frequency of the shaking, the distance from the fault
- The regional geology and site conditions

To estimate the maximum level of shaking expected during the lifetime of a building. Constructing accurate hazard maps is a challenge and remains the focus of much Geoscience research.

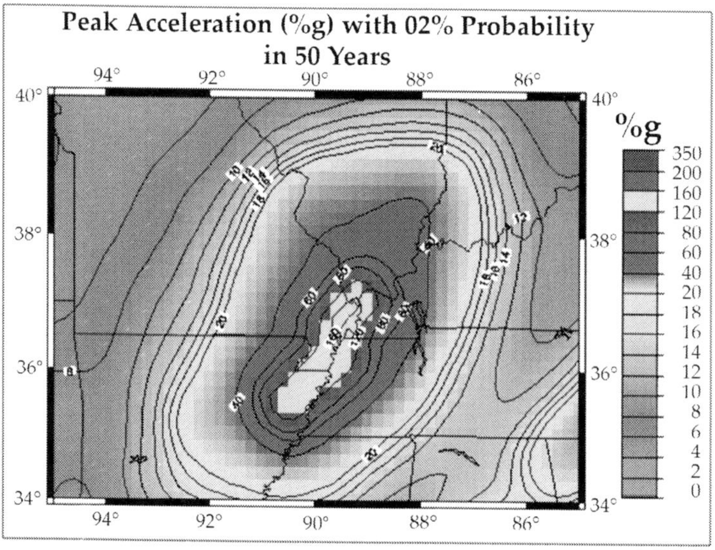

(Courtesy of Dr. Robert Herrmann, Saint Louis University)

Strengthening of Structures against Earthquake Vibrations

We have two methodologies for preparing structures for quakes: either secure the building parts (dividers, floors, establishment, and so forth.) together and have the whole structure carry on as a solitary solid unit that moves with the ground, or you develop a solid and adaptable structure that misshapes yet doesn't break and retains a portion of the shaking vitality. Either approach can be costly so we can't assemble every one of our structures to withstand the biggest conceivable quake. We should make bargains and acknowledge some hazard (this is similar to the dangers that we acknowledge each day, driving on a road, flying in a plane, living in surge inclined locales, tornado "back road", typhoon inclined areas, and so on.).

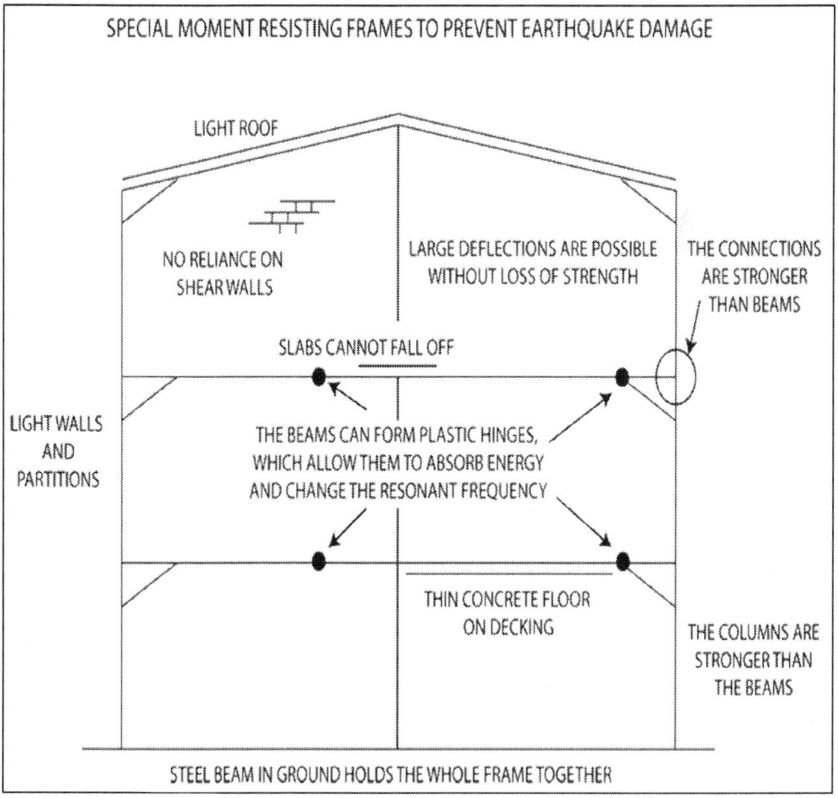

Figure 4.5: A Layout of Building showing Techniques to Strengthen the Structure

We require distinctive levels of resistance for various classes of structures. Basic structures, for example, healing centers, control, water-treatment, and concoction plants, dams, and so on must survive the shaking, as well as must stay in operation. These structures require the biggest venture of assets to guarantee that they can give administrations taking after a seismic tremor.

More general requirements for other structure include having our buildings

- Sustain little damage in small-to-moderate quakes (M < 5.5)
- Sustain some repairable damage for moderate quakes (5.5 < M < 7.0)
- Not collapse in large earthquakes (M > 7.0)

To protect that we meet these objectives, we can make various strides, starting with attentive and capable arranging and zoning laws. Since we realize that destinations with delicate, water-soaked establishments are inclined to harm, we should oppose expansion on those sites and we ought to surely not put basic structures on such locales, and abstain from expanding on these destinations at all if conceivable. On the off chance that this is impractical, we should attempt to conserve the delicate residue before the developing or stay the structure in the cellar. We can find a way to reinforce structures including utilizing steel outline development, satisfactorily securing the structure to the ground through a strong establishment, consolidating shear dividers and additionally cross-supporting into the structure, or more complex methodologies, for example, utilizing elastic or steel cushions to seclude the structure from the shaking.

Figure 4.6: A Photo of Building Collage Due To Low Strengthening During Earthquake

We have talked above seismic waves and how they vibrate the ground which can lead straightforwardly to the fall of structures. There are other, auxiliary impacts that are brought on by quakes, frequently a consequence of solid shaking. A basic case normal in numerous tremors are avalanches. The shaking makes locales of the stone and soil slide downhill. A similar material would in the long run come up short with expanded time, however seismic tremors trigger many slides that do much piece of harm.

FAULT RUPTURE

At the point when the surface blame crack crosses a structure, it puts vast flat or vertical differential removal requests proportional to the blame uprooting over the blame line. There is a bigger shot that a blame crosses a long structure, for example, an interstate, railroad, a viaduct or a pipeline framework.

Figure 4.7: Left Uneven distribution of damage in buildings located in the same area, subjected to the same ground motion. **Right** Totally collapsed buildings, which killed their occupants

Figure 4.8: Fault rupture crossing a railway track (**left**) and a viaduct (**right**)

A blame intersection a railroad curves the rails. This sort of harm is effortlessly repairable since a railroad track is an adaptable structure and the distorted part can be supplanted without much trouble. Nonetheless, when a blame crosses a stiffer structure, for example, a solid viaduct, then it causes perpetual harm. Blame burst amid the 1999 Marmara tremor crossed the Bolu Viaduct on the Ankara-Istanbul motorway (Fig.4.8, right board) with a diagonal point of frequency (Fig. 4.8, left board) and brought on partition of two adjoining docks by around 1.5 m(Hu, Liu, & Dong, 1996).

Figure 4.9: Fault trace running across the Bolu Viaduct piers (left) and the damage caused at the deck structure due to static differential movement of the piers along the ruptured fault

The deck structure above couldn't suit this even removal distinction, and the deck braces practically tumbled off the dock tops (Fig.4.9, right board). The whole deck was disassembled after the seismic tremor and reproduced, which kept the motorway shut for a long time. The fundamental purpose behind this harm was the misallocated blame and the wrong estimation of expected blame uprooting amid outline.

GEOTECHNICAL DEFORMATIONS

Solid ground shaking causes settlements in soil stores which are not compacted adequately. Roadways and railroads developed over stored soil layers are commonplace cases of such ground accidents instigated by extreme ground shaking. Ground disappointment in two paths interstates amid the 2010 Maule, Chile quake is appeared in the left board of Fig.4.10. Amid a similar seismic tremor, a railroad connection project settled broadly and snuck by the railroad tracks, consequently the tracks suspended noticeably all around (Fig.4.10, right board).

Soaked soil stores lose their quality under cyclic burdens driven by solid ground shaking and they stream as a plastic material. This is named as liquefaction. Bearing limit of the melted soil decreases beneath the static gravity stresses forced by the structures above, which thus result in vast vertical disfigurements. Such structures typically sink into the dirt and tilt with an inflexible body movement amid the seismic tremor (Fig.4.11). Ground increasing speeds in condensed soils are typically very low and don't prompt noteworthy auxiliary harm. Subsequently, liquefaction generally does not bring about existence misfortunes or wounds, but rather it causes huge monetary misfortunes because of tilted or uprooted structures. At times, broad ground relocations may prompt halfway crumple, loss of solidness and critical harm

Figure 4.10: Ground failure in a highway (left panel) and extensive settlement in a railroad bridge abutment (right panel) during the 2010 Maule, Chile earthquake (EERI Earthquake Photo Gallery)

Figure 4.11: Tilted buildings on liquefied soils in Adapazarı during the 1999 Marmara earthquake

Reasonable assessment of liquefaction potential could be surveyed from field tests, for example, Standard Penetration Test (SPT) and Cone Penetration Test (CPT). The general thought behind these assessment plans is to look at and evaluate the cyclic resistance proportion, (CRR) with cyclic anxiety proportion (CSR) inside a deterministic or probabilistic system.

Figure 4.12: Lateral spreading in the Olas development landscaping during the 2010 Maule, Chile earthquake (left panel EERI Earthquake Photo Gallery) and buildings dragged into the sea due to lateral spreading in Gölcük during the 1999 Marmara earthquake (right panel)

At the point when soil inclines are subjected to solid ground shaking, they may misshape along the side in an incremental shape as appeared in the left board of Fig. 4.12. This wonder is called parallel spreading. Melted soils spreading along the side can drag the structures resting above them, in some cases into the ocean at a shoreline (Fig.4.12, right board). Its misshaping instrument can be clarified as aggregate lingering uprooting seen along the down-slope bearing amid seismic excitation. Degree of horizontal spreading might be evaluated by making utilization of prescient strategies, rearranged diagnostic models, or numerical (limited components, limited distinction, and so forth.) techniques. Unmistakably, exactness and precision are relied upon to increment from prescient techniques to limited component strategies while time and exertion utilization additionally increment. Geotechnical misshapenness and resulting ground disappointments because of ground shaking amid seismic tremors are concentrated under a different branch of quake building, called Geotechnical Earthquake Engineering. Seismically prompted geotechnical misshapenness are presented in this area quickly. The examination of geotechnical misshapenness under seismic impacts are past the extent of this book.

SOIL LIQUEFACTION

At the point when silt with a high-water substance is subjected to delayed shaking, the weight of the water held in pores in the residue steadily increments in the end, the dregs lose all strong quality and start to act as though they were fluids.

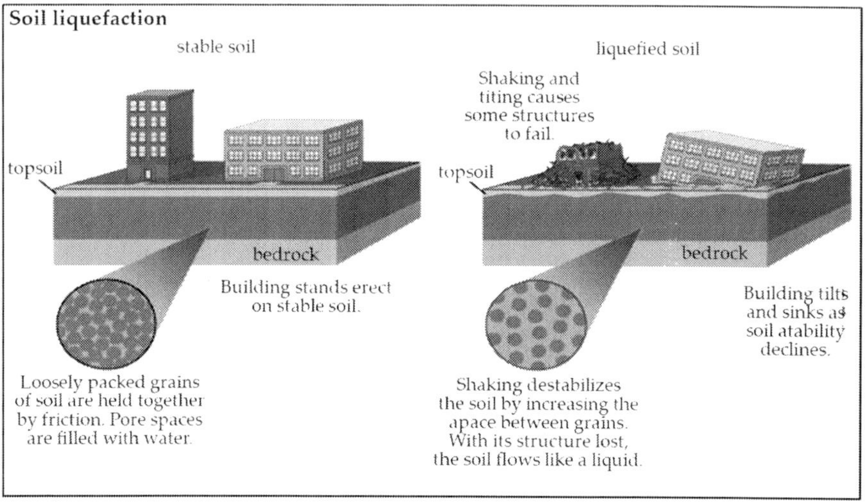

Figure 4.13: The qualities of stable soil compared with those of liquefied soil.

Building and different structures sink into the ground or topple and covered tanks and different cavities ascend to the surface. This is known as liquefaction. Liquefaction occurred amid the quake of 1692 in Jamaica and was in charge of the obliteration of the town of Port Royal.

In the course of recent decades, many Parts of the Eastern Caribbean have turned out to be progressively powerless against liquefaction as a result of the expanded utilization of recovered land for urban advancement.

DIFFERENTIAL SETTLEMENT

In some cases, only part of the foundation is affected by ground failure, or part of the foundation is affected to a greater extent than other parts. This kind of effect is called differential settlement and can cause much more severe damage to a building than uniform or tilt settlement. For instance, one corner of a slab foundation may experience settlement while the rest of the foundation does not. The resulting change in foundation settlement can induce higher levels of stress in the floor, wall and roof members.

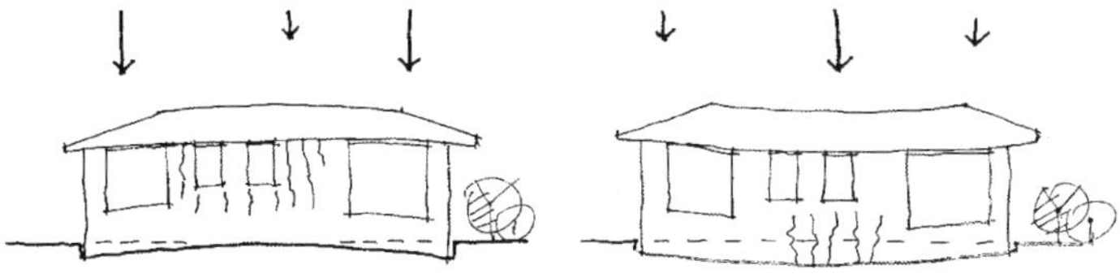

Figure 4.14a: Demonstration of Settlement

Differential settlement that causes the edges of the foundation to settle further than the centre is known as hogging (left). Where it causes the centre to settle further than the edges, it is known as sagging or dishing (right).

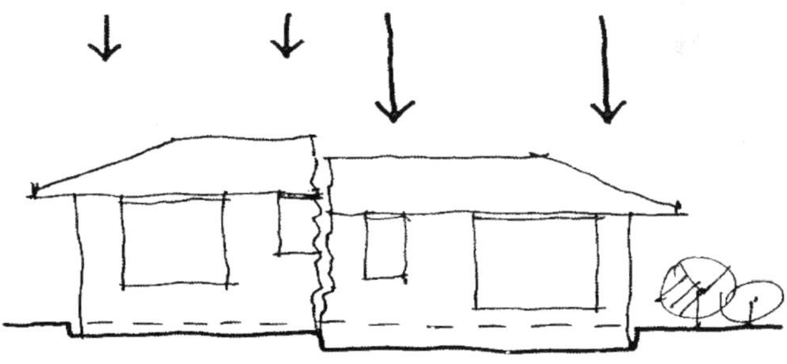

Figure 4.14b: *Demonstration of Settlement*

Differential settlement can cause an abrupt change in level or cause all corners of the foundation to settle by different amounts.

SEISMIC LOADING

The resultant parallel compel or seismic load is spoken to by the drive F as appeared in Fig. The constrain F is unmistakably not quite the same as the dead, live, snow, wind, and effect loads. The even ground movement activity is like the impact of a flat constrain following up on the building, consequently the expression "Seismic Load." As the base of the building moves in a to a great degree confused way, idleness strengths are made all through the mass of the building and its con-tents. It is these reversible strengths that make the building move and manage harm or crumple. Extra vertical load impact is brought on pillars and sections because of vertical vibrations. Being reversible, at specific moments of time the successful load is expanded, at others it is diminished. The seismic tremor burdens are dynamic and difficult to anticipate exactly ahead of time, since each quake displays diverse qualities.

- The following equivalent minimum total lateral force is, used for seismic design:

$$F = S.F_s.I.C.W$$

Where S, F_s, I, C and W are the factors affecting seismic load, which will be explained in the following section.

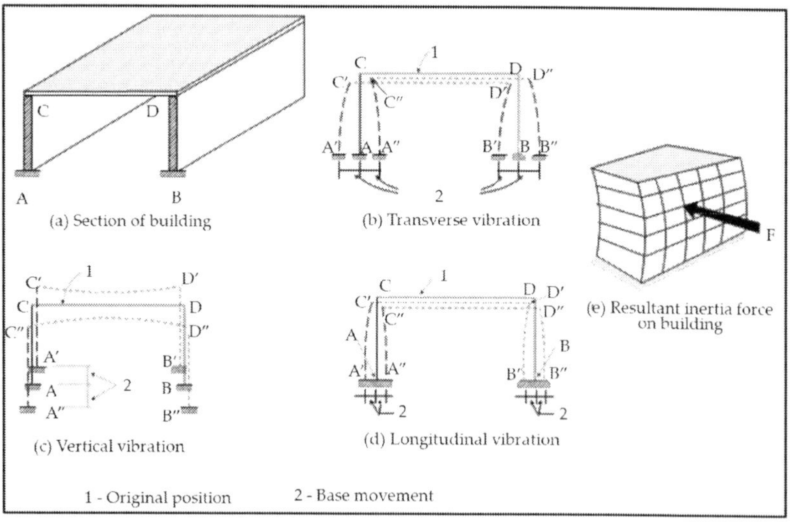

Figure 4.15: Effects of Seismic Loading on Structures

Factors affecting Seismic Loading

The seismic tremor zone consider S relies on the ground force of the earth-shudder. The estimation of S generally is plotted on maps as far as seismic power isolines or most extreme increasing speed isolines. Clearly, the higher the power or increasing speed, the bigger will be the seismic compel. The dirt establishment calculate Fs relies on the proportion of key flexible time of vibration of a working toward the path under thought and the trademark site period. Accordingly, Fs is a numerical coefficient for site-building reverberation. The inhabitance significance or peril figure I relies on the utilization of the building. The higher the significance or bigger the peril brought on by the disappointment of the building, the more noteworthy the estimation of the variable I. The C is a component relying upon the firm ness and damping of the structure. Bigger the firmness for given mass, shorter the basic time of vibration of the structure and bigger the estimation of C. Damping is the vitality dissemination property of the building; bigger the damping, littler the estimation of C. The W is the aggregate weight of the super-structure of a building including its con-tents. The inactivity powers are relative to the mass of the building and just that piece of the stacking activity that has mass will offer ascent to seismic compel on the building. Accordingly, the lighter the material, the littler will be the seismic drive.

TSUNAMIS

An occasionally sensational by-result of specific sorts of tremors are tidal waves. Tidal wave is a Japanese expression that signifies "harbor wave". Waves are much of the time mistaken for tsunamis, yet they don't have anything to do with the tides, they are the consequence of a sudden vertical counterbalanced in the sea depths brought about by quakes, submarine avalanches, and volcanic disfigurement. In 1883 the volcanic emission of Krakatoa brought about the crumple of a caldera that started a torrent which executed 36,000 individuals on adjacent islands. On June 25, 1896, a seismic tremor off the Japanese drift produced a tidal wave that hit the shore with wave statures extending from 10 to 100 feet. As the angling armadas come back to shore taking after an overnight outing they discovered their towns demolished and 22,000 people dead. In the most recent century more than 50,000 individuals have kicked the bucket thus of waves.

Tsunami Initiation

A sudden balance changes the rise of the sea and starts a water wave that ventures outward from the locale of ocean bottom interruption. Torrents can traverse the sea and vast seismic tremors in Alaska and Chile have created waves that brought on harm and passings in locales as far away as California, Hawaii and Japan.

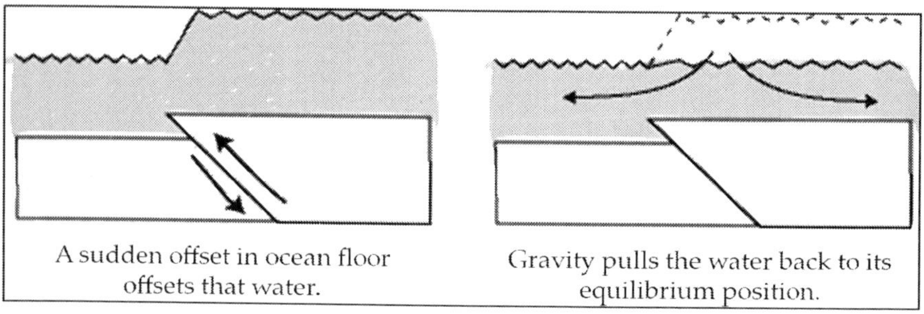

Figure 4.16a: A Pictorial Representation of Tsunami Initiation

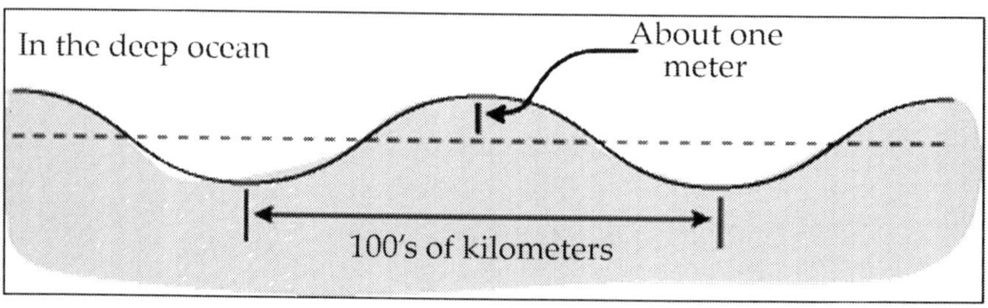

Figure 4.16b: A Pictorial Representation of Tsunami wave

The speed of this wave relies on upon the sea profundity and is normally about as quick as a business traveler fly (around 0.2 km/s or 712 km/hr). This is moderately ease back contrasted with seismic waves, so we are frequently cautioned to the threats of the tidal wave by the shaking before the wave arrives. The inconvenience is that an opportunity to respond is not long in districts near the seismic tremor that brought on the tidal wave.

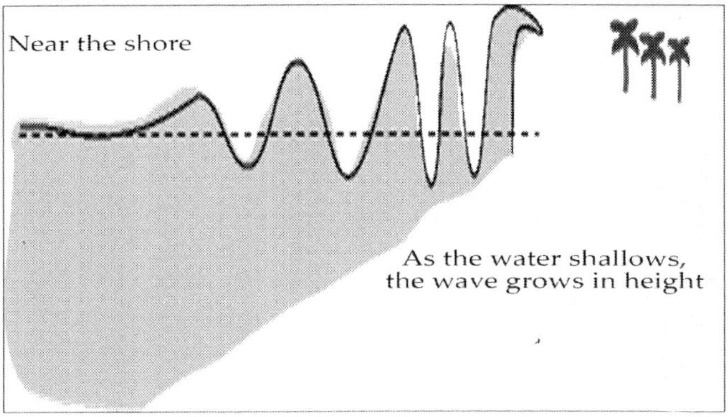

Figure 4.16c: Tsunami Reaching Landmass

Regular statures for vast tidal waves are on the request of 10s of meters and a couple have moved toward 90 meters (around 300 feet). These waves are commonly more destroying to the seaside locale than the shaking of the seismic tremor that brought about the wave. Indeed, even the more typical tidal waves of around 10-20 meters can "wipe clean" beach front groups.

Fatal waves happen about every one to two years and they have now and again slaughtered a huge number of individuals. In 1992-93 three expansive torrents occurred: one in Japan, Indonesia, and Nicaragua. All struck during the evening and crushed the nearby groups.

FLOODS

A surge is a flood of any measure of water that achieves arrive. Surges happen as a rule when the volume of water inside a waterway, for example, a stream or lake, surpasses the aggregate limit of the development, and accordingly a portion of the water streams or sits outside of the ordinary border of the body. Be that as it may, surges might be optional impacts of quakes, if dams are harmed. Tremors may make landslips dam streams, which fall and cause surges, The landscape beneath the Sarez Lake in Tajikistan is in peril of cataclysmic surge if the avalanche dam shaped by the seismic tremor, known as the Usoi Dam, were to fizzle amid a future quake. Affect projections propose the surge could influence approximately 5 million individuals.

Figure 4.17: A flood covering domestic area due to earthquake

HUMAN IMPACTS

A seismic tremor may bring about harm and death toll, street and scaffold harm, general property harm, and fall or destabilization (possibly prompting future crumple) of structures. The outcome may bring sickness, absence of fundamental necessities, mental results, for example, freeze assaults, wretchedness to survivors, and higher protection premiums.

PERFORMANCE OF DIFFERENT STRUCTURES DURING EARTHQUAKE

Different types of buildings suffer different degrees of damage during earthquakes.

Mud and Adobe Houses

Unburnt sun-dried blocks laid in mud mortar are called adobe development. Mud houses are the conventional development for poor and most reasonable in perspective of their underlying cost, simple accessibility, low-level expertise for development and amazing protection against warmth and frosty. More than 100 million individuals in India live in these sorts of houses. There are various cases of the fall of such structures in 1906 Assam, 1948 Ashkhabad, 1960 Agadir, 1966 Tashkent, 1967 Koyna, 1975 Kinnaur, 1979 Indo-Nepal, 1980 Jammu and Kashmir and 1982 Dhamar seismic tremors. These edifices are extremely feeble in shear, strain and pressure. Detachment of dividers at corner and intersections happens effortlessly under ground shaking. The splits go through the poor joints. After the dividers crumble, either because of twisting or shearing in mix with the compressive burdens, the entire house crashes down. Broad harm was seen amid tremor particularly in the event that it happens after a precipitation, (Chandra & Krishna, 1983).

Figure (4.18): A mud house destroyed by earthquake

Better execution occurs by blending the mud with earth to give it a durable quality. The blending of straw enhances the rigidity. Covering the external divider with waterproof substance, for example, bitumen enhances against weathering. The quality of mud dividers can be enhanced altogether by spilt bamboo or timber fortification. Timber edge or even timber runners at lintel level with vertical individuals at corners additionally enhances its imperviousness to parallel strengths which has been seen amid the quakes.

Masonry Buildings

Masonry buildings of brick and stone are superior with respect to durability, fire resistance, heat resistance and formative effects. Masonry buildings consist of various material and sizes

- Large block (block size >50 cm)-concrete blocks, rock blocks or lime stones;
- concrete brick-solid and hollow;
- Natural stone masonry. Because of its easy availability, economic reasons and the merits mentioned above this type of construction are widely used.

In very remote areas in the Himalayas, buildings are constructed of stacks of random rock pieces without any mortar. The majority of new construction use mud mortar, however, few use cement mortar also.

Causes of failure of masonry buildings:

These structures are substantial and pull in extensive idleness strengths. Unreinforced brick work dividers are feeble against strain (Horizontal strengths) and shear, and in this manner, perform fairly poor amid seismic tremors. These structures have huge in plane unbending nature and along these lines have low eras of vibration, which brings about expansive seismic constrain. These structures break apart and given way in light of absence of honesty. The absence of basic respectability could be because of absence of "through" stones, nonappearance of holding between cross dividers, nonattendance of stomach activity of rooftops and absence of box light activity.

Common type of damage in masonry building:

Every one of them experiences serious harm bringing about entire fall and accident in a store of stones. The idleness drives because of rooftop or floor is transmitted to the highest point of the dividers and if the roofing material is shamefully fixing to the divider, it will be removed. The feeble rooftop bolster association is the reason for partition of rooftop from the support and prompts finish crumple. The disappointment of base harmony of rooftop truss may bring about total crumple of truss and in addition the entire building.

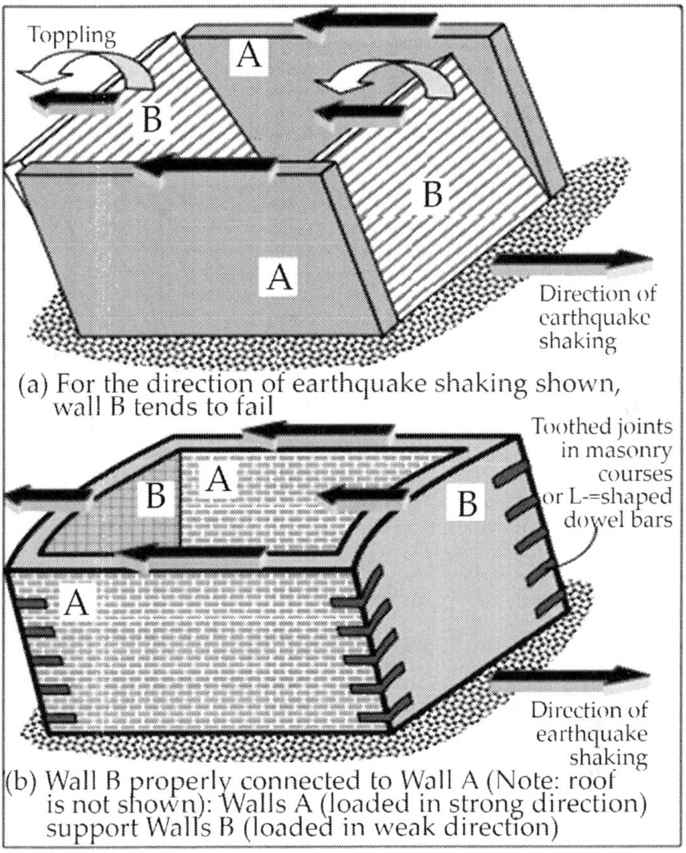

Figure (4.19): Forces acting on masonry structure due to earthquake

On the off chance that the rooftop/floor material is legitimately fixing to the top dividers making it shear of corner to corner toward the path movement through the bedding joints. The breaks more often than not start at the sides of the openings. The disappointment of wharf happens because of consolidated activity of flexure and shear. Close vertical breaks close corner divider joint happen showing detachments of dividers. For movement opposite to the dividers, the bowing minute at the finishes bring about splitting and detachment of the dividers because of poor holding. By and large peak end divider breakdown. Because of vast latency strengths following up on the dividers, the Wythe of brick work is either swell outward or internal. The falling without end of a large portion of the divider thickness on the lump side is basic component. The holding stone is observed to be compelling as in Jammu Kashmir seismic tremor of August 24, 1980. Unreinforced dressed rubble stone work (DRM) has demonstrated marginally preferable execution over irregular rubble workmanship. The most widely recognized harm is because of splits in the dividers. The stone work with lower unit mass and more prominent bond quality shows better execution. The unreinforced stone work generally speaking ought to be evaded as a development material beyond what many would consider possible in seismic territory.

Reinforced Masonry Buildings

Strengthened stone work structures have withstood seismic tremors well, without apparent harm. For level twisting, an intense part fit for taking bowing if found to perform better amid tremors. In the event that the corner areas or opening are strengthened with steel bars considerably more noteworthy quality is accomplished. Indeed, even dry stuffed stone brick work divider with consistent lintel band over openings and cross dividers did not experience any harm.

Brick-R.C Frame Buildings:

This type of building consists of RC frame structures and brick lay in cement mortar as infill. This type of construction is suitable in seismic areas.

Causes of failure of RC frame buildings: The disappointments are because of mostly absence of good outline of bars/sections outline activity and establishment. Low quality of development insufficient enumerating or laying of support in different segments especially at joints and in segments/shafts for malleability. Lacking stomach activity of rooftop and floors. Deficient treatment of brick work dividers.

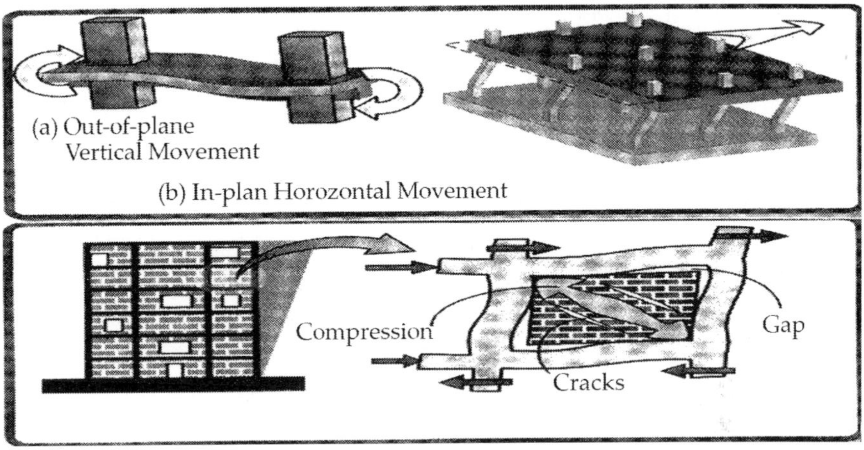

Figure (4.19): effect of lateral focus on columns and beams

Common type of damage in RC frame buildings:

The harm is for the most part because of disappointment of infill, or disappointment of sections or bars. Spalling of cement in segments. Splitting or clasping because of intemperate twisting joined with dead load may harm the segment. The clasping of sections are noteworthy when the segments are slim and the dispersing of stirrup in the segment is substantial. Serious break happens close inflexible joints of edge because of shearing activity, which may prompt finish fall. The differential settlement likewise causes extreme minutes in the casing and may prompt disappointment. Plan of casing ought to be with the end goal that the plastic pivot is bound to bar just, on the grounds that bar disappointment is less harming than the regular disappointment.

Wooden Buildings

This is additionally most normal sort of development in ranges of high seismicity. It is additionally most appropriate material for seismic tremor safe development because of its light weight and shear quality over the grains as seen in 1933 Long shoreline, 1952 Kern nation, 1963 Skopje, and 1964 Anchorage quake.

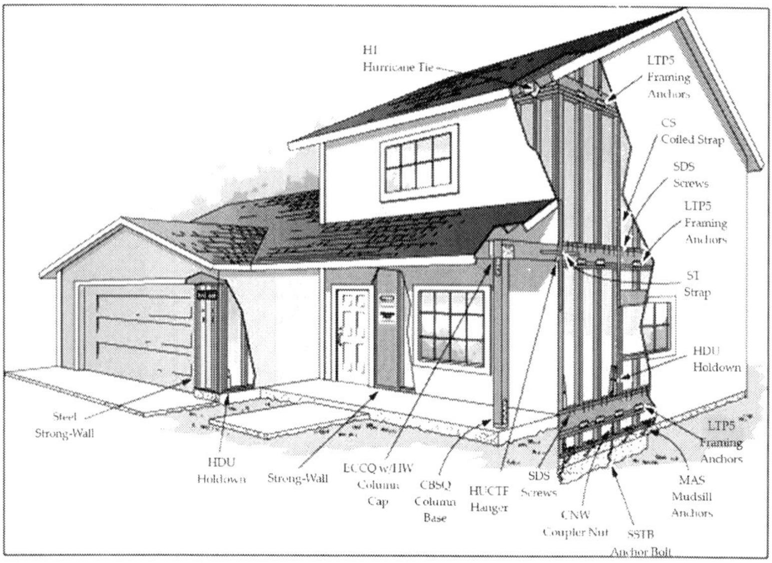

Figure (4.20): Components of Wooden building designed against earthquake

However, amid off-Tokachi tremor (1968), more than 4,000 wooden structures were either absolutely somewhat harmed. Moreover, there were disappointment because of sliding and collapsing because of delicate quality of ground. The primary reason of disappointment was its low unbending nature joints, which goes about as a pivot. Disappointment is additionally because of crumbling of wood with section of time. Wood outlines without dividers have no safe against even strengths. Safe is most elevated for corner to corner propped divider. Structures with askew supporting in both vertical and level plane perform much better.

The customary wood outline Ikra development of Assam and places of Nicobars established on wooden heaps isolated from ground have performed extremely well amid tremors. Wood houses are for the most part reasonable up to two stories.

Reinforced Concrete Buildings:

This kind of development comprises of shear dividers and casings of cement. Considerable harm to strengthened solid structures was found in the Kanto (1923) seismic tremor. Later in Niigata (1964), Off-Tokachi (1968) and Venezuela (1967) seismic tremor it endured substantial harms. The harm to strengthened solid structures might be isolated extensively into vibratory disappointment and tilting or uneven settlement. At the point when a fortified solid building is developed on similarly hard ground vibratory disappointment is seen, while on delicate ground tilting, uneven settlement or sinking is watched.

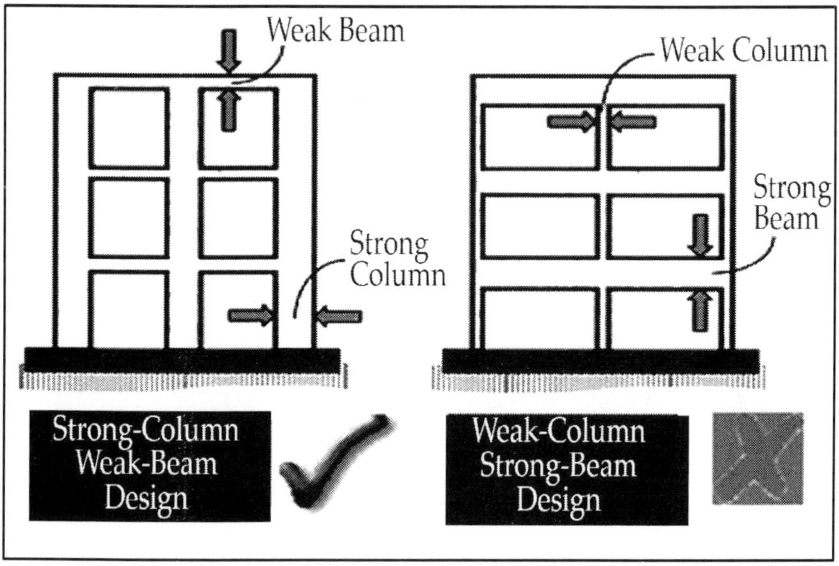

Figure (4.21) Reinforced Concrete Building Design

In the event of vibratory disappointment, the reasons for harm might be thought to be diverse for each case, yet essentially, the seismic powers, which followed up on a working amid the quake, surpassed the heaps considered in the outline, and the structures did not have satisfactory resistance and flexibility to withstand them. By and large these structures performed well as seen in Skopje (1963) and Kern nation (1952) seismic tremors. The shear dividers are affectionate to be powerful to give satisfactory quality to the structures. Extreme harm to spandrel divider between the vertical openings is watched.

Tilting and singing of fortified solid structures amid quakes were found in the Kanto and Niigata tremors. Most fizzled on the grounds that the dead weights couldn't be upheld after the settling of the ground. Such harm is impossible to miss to structures in delicate ground, the harm ends up noticeably higher in the accompanying request: heap establishment, tangle establishment, constant establishment and autonomous establishment.

The empty solid square structures with steel fortification in chose grout filled cells have indicated great execution. The Precast and pre-focused on fortified solid structures additionally endured extreme harm for the most part on account of poor conduct of joints or backings. The Precast and pre-focused on component when in doubt were not demolished as seen in 1952 Kern nation and 1964 Anchorage seismic tremors.

Steel Skeleton Buildings:

Structures with steel skeleton development contrast enormously as indicated by states of cross areas and technique for association.

Figure (4.10). Steel Skeleton Buildings (bridge)

They might be comprehensively isolated into two assortments, those utilizing supports as seismic tremor safe components and those which are inflexible edge structures. The previous is utilized as a part of low building while the latter is tilized as a part of elevated structures.

At the point when props are utilized as seismic tremor safe components, it is typical to outline so that every even compel will be borne by the supports. This kind of building is by and large light and impact of wind burdens is overwhelming by and large. In any case, there are many cases in which the props have demonstrated softening or locking in which joints have fizzled(Wiegel, 1970). Steel skeleton development, especially the basic sort in which casings are involved shafts and segments comprising of single part H-pillars, is frequently utilized as a part of tall structures. The non-auxiliary harm is normal however none of these building extremely harms as seen in 1906 San Francisco seismic tremor.

Steel and Reinforced Concrete Composite Structures:

Steel and Reinforced Concrete Composite Structures are made out of steel skeleton and fortified cement and have the dynamic qualities of both. It is better concerning imperviousness to fire and wellbeing against clasping when contrasted with steel skeleton. While contrasted with strengthened solid structures it has better malleability in the wake of yielding. As these elements are the properties, which are powerful to make a building seismic tremor safe and are, found to perform better amid quakes (Wiegel, 1970).

5 CHAPTER

RESPONSE OF SIMPLE STRUCTURES TO EARTHQUAKE

GENERAL

The goals of this chapter are to address and bring together meanings of the crucial reaction parameters thought to be most persuasive in basic quake designing, and to highlight the elements affecting these key reaction parameters. The parameters hypothesized in this book to be the essential building squares of understanding and controlling seismic tremor reaction of structures are Stiffness, Strength and Ductility. As displayed in the accompanying section, firmness is the most related parameter in reacting to the necessities of serviceability under the little successive tremor. In a practically equivalent to way, quality is used to control the level of inelasticity under the medium, estimated occasional seismic tremor, thus it maps onto the harm control confine state. At long last, fall avoidance under the vast uncommon quake is most influenced by pliability, in this manner finishing the peril, confine state, reaction parameter ternions examined before. The material in this part is introduced in an entirely various leveled structure of material, segment, part, association and framework attributes most compelling in influencing firmness, quality and pliability. The section closes with a treatment of the two essential amounts of over quality and damping, which are important to the three central parameters talked about above.

CONCEPTUAL FRAMEWORK

Definitions

So as to set up a typical classification and in acknowledgment of the plenty of clashing definitions in the writing, bland and thorough meanings of the fundamental terms utilized as a part of this book are given in this. Concentrate is put on the three reaction qualities utilized from this point forward as the most critical parameters that portray the conduct of structures and their establishments when subjected to seismic tremors. These are firmness, quality (or limit) and flexibility. Before characterizing the three amounts, it is educational to repeat the meaning of two more major amounts, to be specific "activity " and "misshapening " . The previous is utilized as a part of this book to show push resultants of different types, while the last is utilized to demonstrate strain resultants. The three amounts of solidness, quality and pliability are dealt with in detail in ensuing areas of this section, and are concisely characterized from now on so as to allow a sane dialog of the calculated system for the entire content.

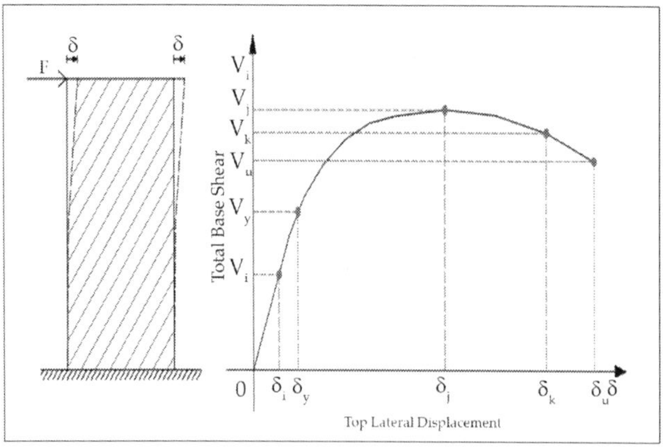

Figure 5.1: Typical response curve for structural systems subjected to horizontal loads

Stiffness is the ability of a component or an assembly of components to resist deformations when subjected to actions, as shown in Figure 5.1. It is expressed as the ratio between action and deformation at a given level of either of the two quantities and the corresponding value of the other. Therefore, stiffness is not a constant value. In Figure 5.1, Ki is the stiffness at a required deformation δi and corresponds to force resistance Vi. If increments or first derivatives of actions and deformations are used, the ensuing stiffness is the tangent value. If total actions and deformations are used, the ensuing stiffness is the secant value.

Strength is the capacity of a component or an assembly of components for load resistance at a given response station. It is also not a constant value, as shown in Figure 5.1. In this book, the term 'capacity' is preferred to the term 'strength' to represent both action resistance and the ability to endure deformation, or deformation capacity. In the figure, Vj and Vk are the force capacities corresponding to δj and δ k, respectively. V y, referred to as the yield strength, corresponds to the yield displacement δ y, which is required for ductility calculations.

Ductility is the ability of a component or an assembly of components to deform beyond the elastic limit, as shown in Figure 5.1, and is expressed as the ratio between a maximum value of a deformation quantity and the same quantity at the yield limit state. In the figure, the displacement ductility μ is the ratio between the maximum or ultimate displacement δu and the yield displacement δy.

Demand is the action or deformation imposed on a component or an assembly of components when subjected to earthquake ground motion. This demand is not constant. It continuously varies as the structural characteristics vary during inelastic response. It also varies with the characteristics of the input motion.

Supply is the action or deformation capacity of a component or an assembly of components when subjected to earthquake ground motion. Therefore, the supply represents the response of the structure to the demand. It may continuously vary as the structural characteristics change during inelastic response. It also varies with the characteristics of the input motion. For inelastic systems, demand and supply are coupled(Elnashai & Di Sarno, 2008).

Strength, versus Ductility, Based Response

Traditional force, based seismic design has relied on force capacity to resist the earthquake effects expressed as a set of horizontal actions defined as a proportion of the weight of the structure. In the past 20 to 30 years, there has been a tendency to substitute ductility (or inelastic deformation capacity) for strength (or force capacity).

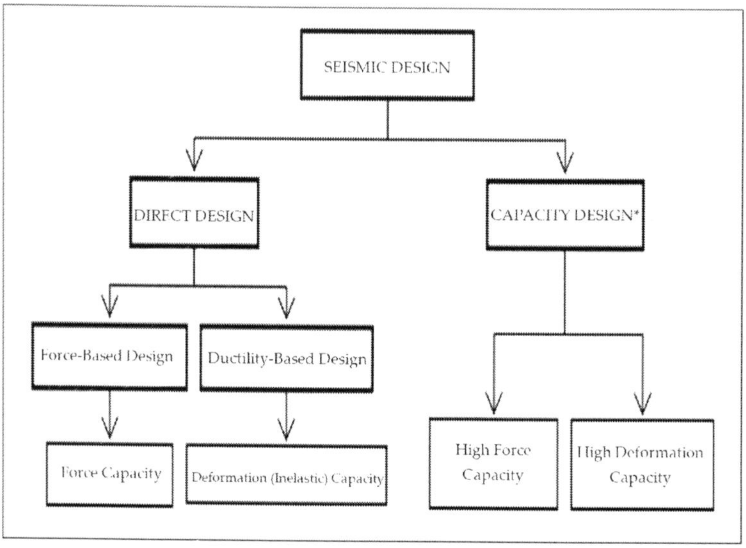

Figure 5.2: Different approaches to seismic design

The last approach was produced in acknowledgment of the considerable instability related with assessing seismic request. A flexibility, planned structure is altogether less delicate to surprising increment in the compel request forced on it than its quality, composed partner. By and large, malleability, based structures are lighter and utilize less material, however more workmanship. An alternate method for dimensioning tremor, safe structures is 'limit configuration' (Figure 5.2). Limit configuration utilizes a blend of individuals with high load limit and individuals with high inelastic distortion ability to enhance the reaction of the auxiliary framework. This is accomplished by recognizing a disappointment component, the individuals and districts in charge of its advancement, and giving these individuals and locales with satisfactory flexibility. In parallel, whatever is left of the structure is secured by giving it sufficient quality to guarantee about versatile conduct. The inverse of 'limit configuration' is 'immediate plan', which is the dimensioning of individual segments to oppose the privately assessed activities with no due thought to the activity, redistribution impacts in the framework in general. Coordinate plan can be either malleability, based or quality, based. Limit plan, then again, depends on both quality and pliability of segments.

The difference between direct and capacity design is depicted in Figure 5.3. The maximum effects of both horizontal and vertical loads are computed through structural analysis. In the direct approach, all design actions are the 'applied' quantities, calculated from the combination of static and seismic loads. In capacity design, one set of actions represents the ultimate capacity of the members, regions or mechanisms that are responsible for energy absorption, while the rest of the design actions are calculated to maintain equilibrium. In the figure, Mc12, M_{c11} and M_b are the moments on the two columns and the beam, respectively, evaluated from the applied actions. M_{bmax} is the maximum capacity of the beam, the member responsible for energy absorption in the weak, beam strong, column design approach, taking into account various sources of over strength (e.g. unintentional increase in material properties, rounding, off of member or reinforcement dimensions, post, yield hardening, etc.). Design actions Mc12′ and $M_{c11'}$ are the product of M_{c12} and M_{c11}, and the ratio M_{bmax}/M_b. They are there for 'applied' actions(Elnashai & Di Sarno, 2008).

Member, versus System, Level Consideration

Just as of late has the seismic tremor designing group considered the general framework reaction completely? Traditional seismic plan prescribes the dimensioning of individuals to oppose the activities radiating from basic investigation where the dead and live loads are connected close by a calculated estimation of proportional even tremor activities, as appeared for instance in Figure 5.3 .The communication amongst part and framework in basic quake building is unpredictable, however its comprehension is basic for compelling seismic outline. The various leveled connection amongst nearby and worldwide basic reaction is given in Figure 5.4; it is relevant to the three basic amounts of solidness, quality and pliability.

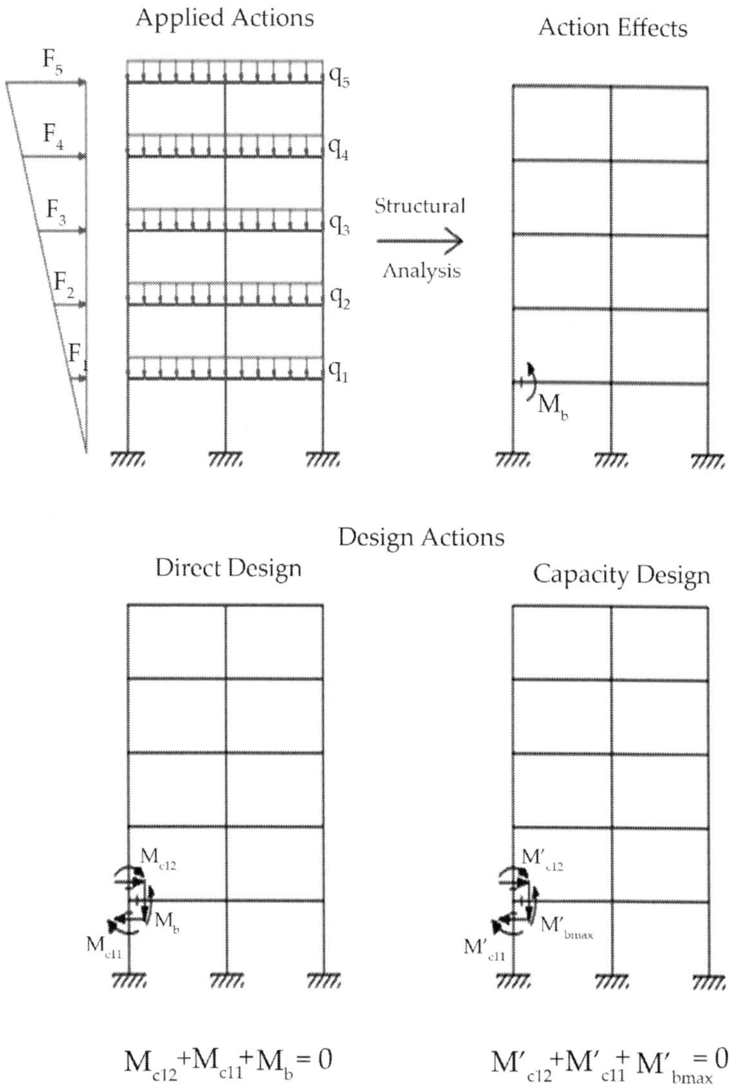

Figure 5.3: Comparison between direct and capacity design

Quantitative expressions connecting neighborhood activity – disfigurement qualities to worldwide reaction amounts must be determined under glorified conditions and with various improving suspicions that point of confinement their extent of utilization. A subjective energy about the neighborhood – worldwide association, and application to particular cases, are fundamental to controlled seismic execution. The applied structure of framework reaction may likewise be utilized to survey seismic request and supply. The yield of one part or more does not highlight in the framework activity – disfigurement reaction, and thus is not really a framework constrain state.

The chain framework proposed by Paulay and Priestley (1992) to depict the justification behind the limit outline reasoning is successful as a reason for clarification, as opposed to application, of the idea. An arrangement bind is lacking to describe the perplexing reaction of structures under seismic tremor loads. Systems, e.g. street systems, give a premise to theoretical and pictorial depiction of the seismic conduct of structures (Figure 5.5) and furthermore demonstrate that hindrances between sub-orders are artificial.

The transportation arrange joins parallel and arrangement disappointment modes that may happen in basic frameworks. It likewise gives a visual depiction of other imperative parts of seismic outline, for example, stack way (heading of activity stream), limit (most extreme stream limit of a connection) and plastic redistribution (probability of option courses on account of movement blockage). In the system similarity, the request on the system (seismic tremor activities) might be communicated by the starting point, goal sets forced on the system. The limit of streets to convey activity symbolizes the limit of auxiliary individuals (Hu et al., 1996).

Nature of Seismic Effects

Dissimilar to most different sorts of dynamic activities, seismic tremor impacts are not forced on the structure but rather created by it. Accordingly, two structures established on a similar soil a couple meters separated may need to oblige endlessly unique activity and distortion requests, contingent upon their own mass, firmness, quality and pliability.

It is contended in this that the key amounts are not period and damping, since period is a component of mass and firmness (and quality in the inelastic range) and the primary wellspring of damping in quake building is vitality retention by inelastic misshapening. Setting the "mass" term aside, seismic tremor reaction is influenced in a perplexing way by solidness, quality (or limit) and pliability. In a basic rendition of this mind-boggling issue, firmness directs vibration periods, subsequently enhancement. Changes in firmness cause detuning of the structure and the info movement henceforth additionally influences intensification.

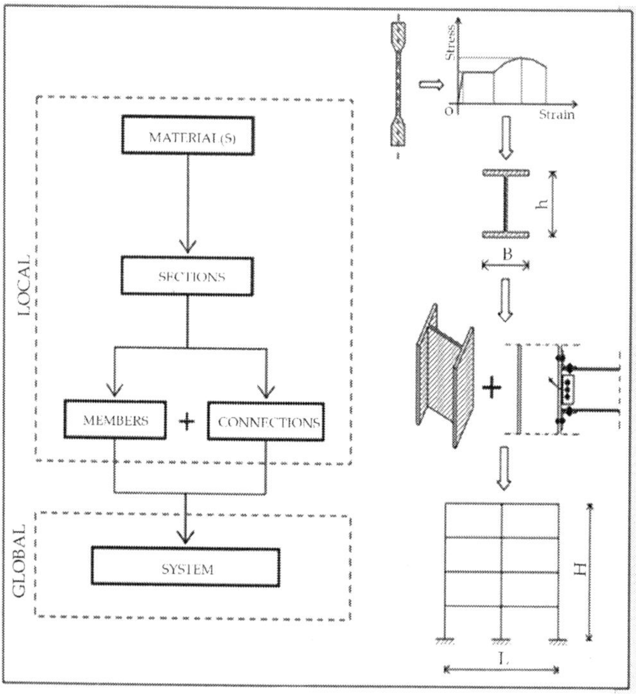

Figure 5.4: Hierarchical relationship between local and global structural response

Figure 5.6 gives a case of two single-level of-flexibility (SDOF) frameworks, one stiffer than the other, yet the dynamic enhancement is with the end goal that the less solid structure (taller wharf) dislodges not as much as the stiffer structure.

Quality points of confinement portray the district where the structure can manage irreversible harm subsequently retains and disperses the seismic activity. Quality cutoff points accordingly prompt the following stage, which is inelastic flexible reaction. Pliability is a vitality sink, in this manner it could be viewed as equal damping. Its impact on basic reaction is, like damping, extension of the reaction time frame and diminishment in the vibration sufficiency.

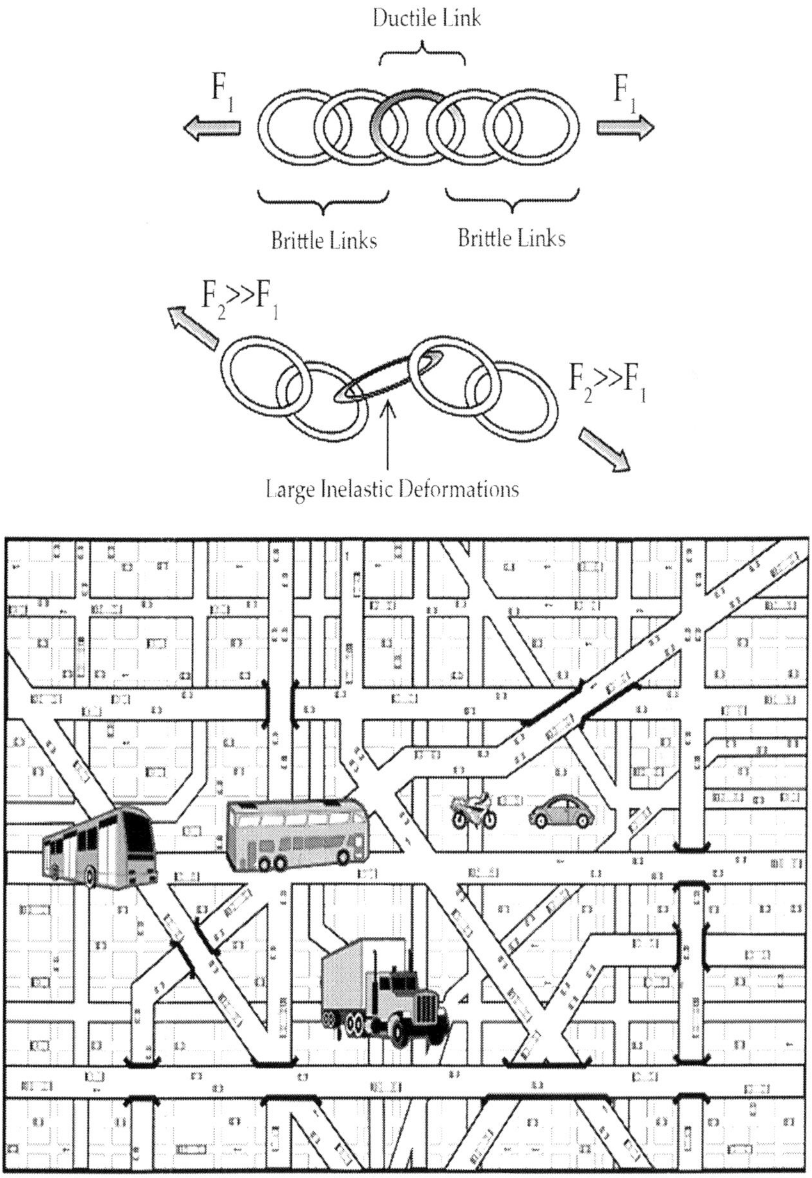

Figure 5.5: Capacity design analogy: chain (top) versus network system (bottom)

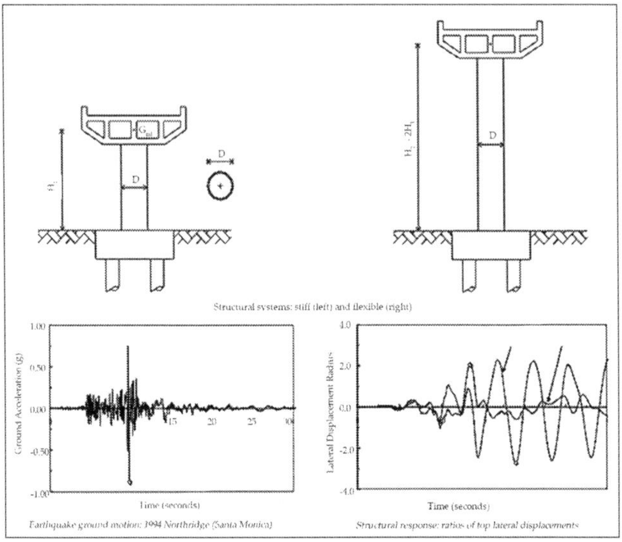

Figure 5.6: Seismic response of structural systems with different dynamic characteristics: stiff and flexible piers

Fundamental Response Quantities

Solidness and quality are not generally related. For a solitary auxiliary part, or a structure that utilizes just a solitary mode basic framework (e.g. outlines just, trusses just or dividers just), they are corresponding. It is, be that as it may, enlightening to investigate situations where they are not corresponding or their consistent of proportionality can be changed. The inspiration for so doing is to pick up a more profound knowledge into the parts of reaction, furnishing engineers with an arrangement of devices that empower fine, tuning of both their comprehension of seismic reaction and their following outline. A straightforward case of the decoupling of firmness and quality is the idea of 'particular intercession' for seismic retrofitting (Elnashai, 1992) as indicated pictorially in Figure 5.7. A structure that was outlined utilizing direct (quality or even malleability) plan where it is required to change it to a limit, composed structure, the main prerequisite is to change the quality dissemination without fundamentally changing the firmness circulation. Different situations are given in the last reference and indicated tentatively to be absolutely sensible (Elnashai and Pinho, 1998).

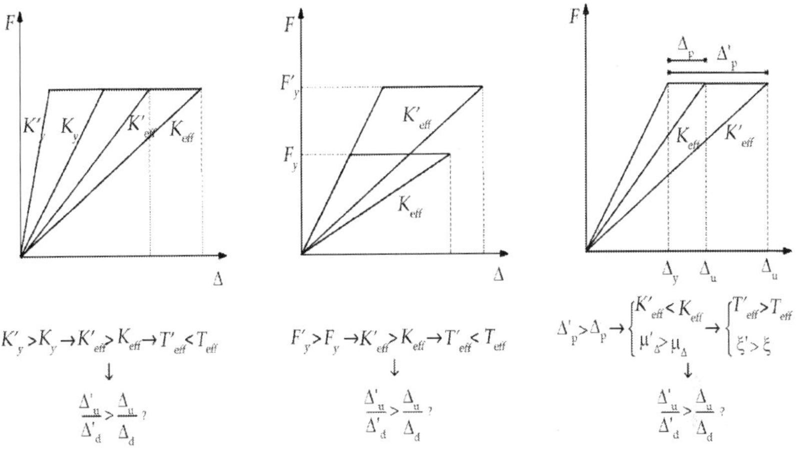

Figure 5.7: Conceptual depiction of the effects of stiffness, strength and ductility variations on system response in seismic retrofitting of structures

Another case is a blended, mode outline, divider fortified cement (RC) structure. Changing the proportion of dividers in a story quickly changes the solidness and quality in an unbalanced way. In this book, the three principal amounts of firmness, quality and flexibility are utilized to clarify issues of seismic reaction of basic frameworks inside a system that is to some degree particular from current patterns(Elnashai & Di Sarno, 2008).

Social and Economic Limit States

Thus, with regards to building up a typical vocabulary through the verbalization of a reasonable structure, nonexclusive cutoff states are talked about concerning the impact of quakes on defenseless groups. At the point when subjected to little quakes, a general public looks for minimal disturbance from harm. This might be viewed as a 'continuous utilize' confine state, and is unmistakably most associated with structures having satisfactory solidness to oppose experiencing vast disfigurements. At the point when subjected to medium quakes, a general public would endure disturbance to its attempts, however would look to limit repair costs. This might be seen as a 'controlled financial misfortune's point of confinement state and is most identified with the structure having satisfactory quality so that the harm is restricted. At long last, when subjected to substantial seismic tremors, a general public would acknowledge intrusion, high financial misfortune, however would try to limit death toll. This is an 'existence wellbeing' constrain state and is most influenced by the flexibility of the structure that empowers it to distort well into the inelastic range, without noteworthy loss of imperviousness to gravity activities. The principal reaction attributes of firmness, quality and pliability are subsequently unmistakably identified with the most essential reaction restrict conditions of proceeded with utilize, harm control and life security (Table 5.1), which are thusly identified with the financial contemplations overseeing the response of groups.

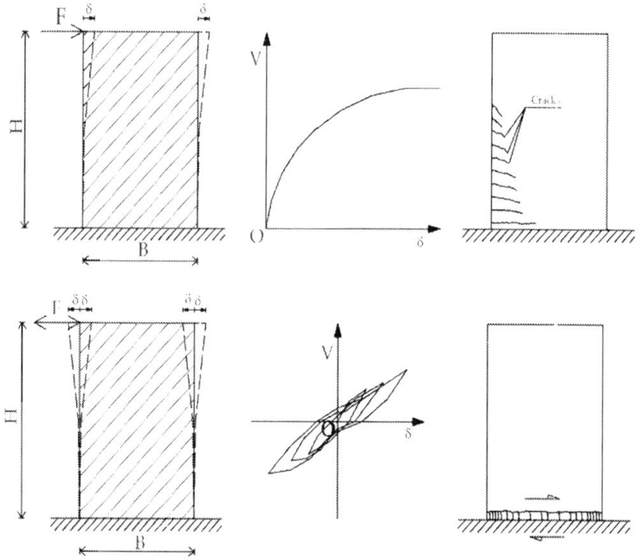

Figure 5.8: Structural RC walls under monotonic or low-cycle cyclic loading (top) and severe cyclic loading (bottom)

The particular qualities given in Table 5.1 might be debated on a building premise. For instance, the quake size related with an arrival period is intensely reliant on the seismo, structural condition under thought. Additionally, more than three cutoff states show up in numerous productions on tremor building (e.g. Bertero, 2000; Bozorgnia and Bertero, 2004, among others). Be that as it may, the above proposed system is powerful and uses central auxiliary reaction qualities that are blandly connected to reaction or execution targets and guide onto social and financial prerequisites. Transitional point of confinement states are valuable in particular cases, and are to a great extent connected with the unique prerequisites of partners(Elnashai & Di Sarno, 2008).

Table 5.1: Relationship between earthquakes, structural characteristics and limit states.

Return period (years)	Earthquake magnitude	Structural characteristics	Engineering limit state	Socio-economic limit state
~75–200	~4.5–5.5	Stiffness	Insignificant damage	Continued operation
~400–500	~5.5–6.5	Strength	Repairable damage	Limited economic loss
~2,000–2,500	~6.5–7.5	Ductility	Collapse prevention	Life loss prevention

6
CHAPTER

DYNAMIC RESPONSE ANALYSIS

INTRODUCTION

The primary driver of harm to structures amid a seismic tremor is their reaction to ground movements which are contribution at the base. So as to assess the conduct of the structure under this kind of stacking condition, the standards of basic flow must be connected to decide the burdens and avoidances, which are produced in the structure. Auxiliary architects know about the examination of structures for static loads in which a heap is connected to the structure and a solitary arrangement is gotten for the subsequent removals and part compels. While considering the examination of structures for dynamic movements, the term dynamic just signifies "time-differing". Thus, the stacking and all parts of the reaction differ with time. This outcomes in conceivable arrangements at every moment amid the time interim under thought. From a building point of view, the most extreme estimations of the basic reaction are typically the ones specifically noteworthy, particularly on account of auxiliary outline.

The reason for this part is to present the standards of basic flow with accentuation on seismic tremor reaction examination. Consideration will at first be centered around the reaction of basic auxiliary frameworks, which can be spoken to regarding a solitary level of flexibility. The ideas created for these frameworks will then be stretched out to incorporate summed up single-level of-flexibility (SDOF) frameworks utilizing the summed up facilitate approach. This improvement thusly prompts the thought of the reaction of structures having numerous degrees of opportunity. At last, ideas and methods utilized as a part of nonlinear dynamic-reaction investigation will be presented.

Amid serious seismic tremors, nonlinear reaction of soil is seen in the broad territory. In the dirt structure connection investigation, in this way, non-linearity in the reaction ought to be considered not just for the structure and soil close to the structures additionally for far field. There are two strategies to manage nonlinear soil structure association issues. One is the immediate strategy in which soil is displayed up to a counterfeit limit. Since the limit reproduces the vast degree of soil just roughly, the limit ought not be put close to the structure, prompting substantial degrees of opportunities. The other strategy is sub-structure technique. In the substructure strategy, structure and adjacent soil with sporadic geometry, material heterogeneity, and nonlinear practices are situated in one substructure, and unbounded soil with normal geometry and straight material properties is dissected in another substructure. All things considered, non-linearity ought to be limited close to the structure. Here, a strategy for treating non-linearity in the inconceivable degree of establishment soil is proposed. An uncommon limited distinction strategy is produced for nonlinear dynamic reaction investigation of semi-unending establishment soil. At to begin with, semi-limitless space is mapped into limited area utilizing unique mapping. At that point, the subsequent conditions are fathomed with the two-stage Lax-Windproof plots on a uniform work of limited contrasts. Material non-linearity can be considered in the whole semi-interminable area. This is the augmentation of the technique, which mapping endless space, into limited area depicted into nonlinear issue.

DYNAMIC EQUILIBRIUM

The essential condition of static balance utilized as a part of the relocation technique for investigation has the frame, where p is the connected compel, k is the firmness resistance, and v is the subsequent removal.

On the off chance that the statically connected compel is currently supplanted by a dynamic or time-fluctuating power p(t), the condition of static harmony ends up plainly one of dynamic balance and has the shape where a dab speaks to separation regarding time.

$$p = kv \tag{6-1}$$

An immediate examination of these two conditions demonstrates that two huge changes, which recognize the static issue from the dynamic issue, were made to Equation 6-1 keeping in mind the end goal to acquire Equation 6-2. To begin with, the connected load and the subsequent reaction are presently elements of time, and consequently Equation 6-2 must be fulfilled at every moment of time amid the time interim under thought. Hence it is generally alluded to as a condition of movement. Also, the time reliance of the removals offers ascend to two extra strengths which oppose the connected compel and have been added to one side hand side.

$$p(t) = m\ddot{v}(t) + c\dot{v}(t) + kv(t) \tag{6-2}$$

The condition of movement speaks to a declaration of Newton's second law of movement, which expresses that a molecule followed up on by a compel (torque) moves so that the time rate of progress of its straight (precise) energy is equivalent to the where the rate of progress of the dislodging as for time, dv/dt, is the speed, and the force is given by the result of the mass and the speed. Review that the mass is equivalent to the weight isolated by the speeding up of gravity. If the mass is constant, Equation 6-3 becomes

$$p(t) = \frac{d}{dt}\left(m\frac{dv}{dt}\right) \tag{6-3}$$

$$p(t) = m\frac{d}{dt}\left(\frac{dv}{dt}\right) = m\ddot{v}(t) \tag{6-4}$$

which states that the force is equal to the product of mass and acceleration. According to d'Alembert's principle, mass develops an inertia force, which is proportional to its acceleration and opposing it. Hence the first term on the right-hand side of Equation 6-2 is called the inertia force; it resists the acceleration of the mass.

Dissipative or damping forces are inferred from the observed fact that oscillations in a structure tend to diminish with time once the time-dependent applied force is removed. These forces are represented by viscous damping forces, that are proportional to the velocity with the constant proportionality referred to as the damping coefficient. The second term on the right-hand side of Equation 6-2 is called the damping force. Inertia forces are the more significant of the two and are a primary distinction between static and dynamic analyses.

It must also be recognized that all structures are subjected to gravity loads such as self-weight (dead load) and occupancy load (live load) in addition to dynamic base motions. In an elastic system, the principle of superposition can be applied, so that the responses to static and dynamic loadings can be considered separately and then combined to obtain the total structural response. However, if the structural behavior becomes nonlinear, the response becomes load-path-dependent and the gravity loads must be considered concurrently with the dynamic base motions.

Under strong earthquake motions, the structure will most likely display nonlinear behavior, which can be caused by material nonlinearity and/or geometric nonlinearity. Material nonlinearity occurs when stresses at certain critical regions in the structure exceed the elastic limit of the material. The equation of dynamic equilibrium for this case has the general form

$$p(t) = m\ddot{v}(t) + c\dot{v}(t) + k(t)v(t) \tag{6-5}$$

in which the stiffness or resistance k is a function of the yield condition in the structure, which in turn is a function of time. Geometric nonlinearity is brought on by the gravity loads following up on the distorted position of the structure. On the off chance that the horizontal relocations are little, this impact, which is frequently alluded to as P-delta, can be disregarded. Be that as it may, if the parallel removals turn out to be vast, this impact must be considered. With a specific end goal to characterize the inactivity drives totally, it is important to consider the increasing speeds of each mass molecule in the structure and the comparing relocations. Such an answer would be restrictively tedious. The examination system can be enormously rearranged if the mass of the structure can be concentrated (lumped) at a limited number of discrete focuses and the dynamic reaction of the structure can be spoken to as far as this predetermined number of relocation segments. The quantity of dislodging segments required to determine the position of the mass focuses is known as the quantity of dynamic degrees of opportunity. The quantity of degrees of flexibility required to acquire a sufficient arrangement will rely on the many-sided quality of the basic framework. For a few structures, a solitary level of flexibility might be adequate, though for others a few hundred degrees of opportunity might be required.

SINGLE-DEGREE-OF-FREEDOM SYSTEMS

Time-Dependent Force

The least complex structure that can be considered for dynamic investigation is a glorified, one-story structure in which the single level of opportunity is the sidelong interpretation at the rooftop level as appeared in Figure 6-1. In this admiration, three imperative presumptions are made. To begin with, the mass is thought to be concentrated (lumped) at the rooftop level. Second, the rooftop framework is thought to be inflexible, and third, the hub misshapening in the sections is disregarded. From these suppositions, it takes after that all horizontal resistance is in the opposing components, for example, sections, dividers, and slanting supports situated between the rooftop and the base. Utilization of these presumptions brings about a discretized structure that can be spoken to as appeared in either Figure 6-lb or 6-1c with a period subordinate constrain connected at the rooftop level. The aggregate firmness k is essentially the whole of the stiffnesses of the opposing components in the story level.

The forces acting on the mass of the structure are shown in Figure 6-1d. Summing the forces acting on the free body results in the following equation of equilibrium, which must be satisfied at each instant of time:

$$f_i + f_d + f_s = p(t) \qquad (6\text{-}6)$$

Where,

fi = inertia force = mu

f_d = damping (dissipative) force= cv&

f_s = elastic restoring force = kv

p(t) = time-dependent applied force

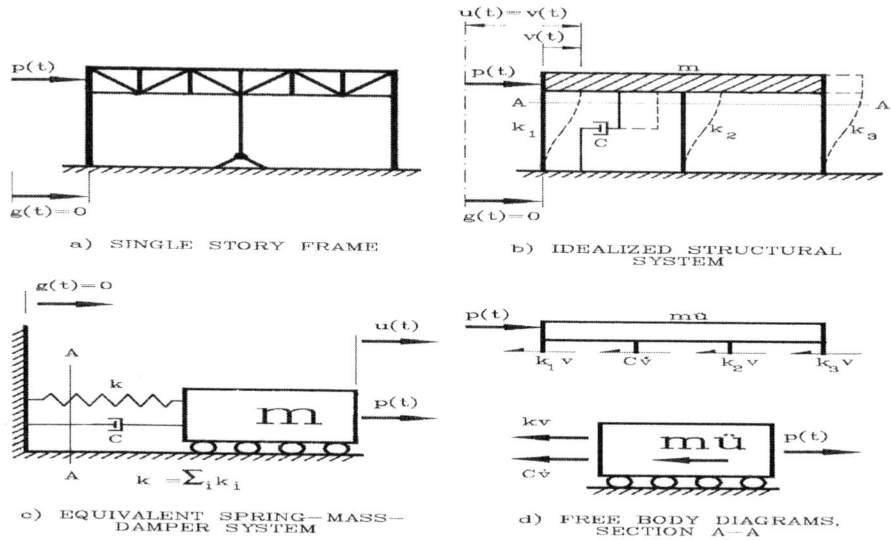

Figure 6.1: Single-degree-of-freedom system subjected to time-dependent force

“u” is the total acceleration of the mass, and v&, v are the velocity and displacement of the mass relative to the base. Writing Equation 4-6 in terms of the physical response parameters results in

mu + cv + kv + p(t) (6-7)

It should be noted that the forces in the damping element and in the resisting elements depend upon the relative velocity and relative displacement, respectively, across the ends of these elements, whereas the inertia force depends upon the total acceleration of the mass. The total acceleration of the mass can be expressed as:

$$\ddot{u}(t) = \ddot{g}(t) + \ddot{v}(t) \qquad (6\text{-}8)$$

where

$\ddot{v}(t)$ = acceleration of the mass relative to the base

$\ddot{g}(t)$ = acceleration of the base

In this case, the base is assumed to be fixed with no motion, and hence g and u(t) v(t).

$$m\ddot{v} + c\dot{v} + kv = p(t) \tag{6-9}$$

When a single-story structure, shown in Figure 6-2a, is subjected to earthquake ground motions, no external dynamic force is applied at the roof level. Instead, the system experiences an acceleration of the base. The effect of this on the idealized structure is shown in Figure 6-2b and 6-2c. Summing the forces shown in Figure 6-2d results in the following equation of dynamic equilibrium:

$$f_i + f_d + f_s = 0 \tag{6-10}$$

Substituting the physical parameters for fi, fd and fs in Equation 6-10 results in an equilibrium equation of the form

$$m\ddot{u} + c\dot{v} + kv = 0 \tag{6-11}$$

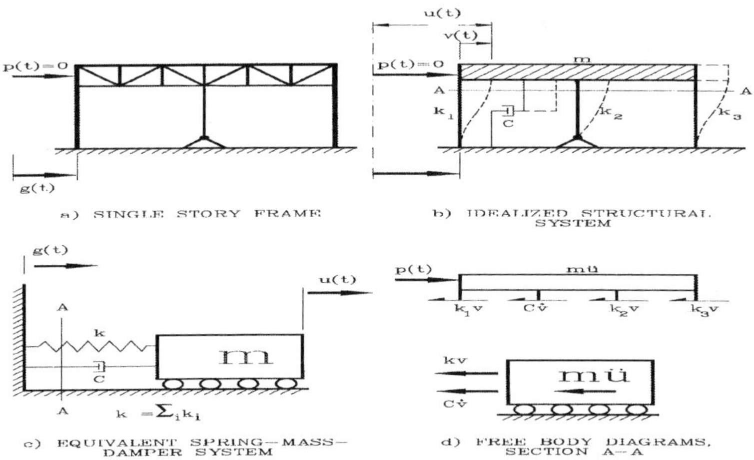

Figure 6.2: Single-degree-of-freedom system subjected to base motion

Hence the equation of motion for a structure subjected to a base motion is similar to that for a structure subjected to a time-dependent force if the base motion is represented as an effective time-dependent force which is equal to the product of the mass and the ground acceleration.

Mass and Stiffness Properties

Most SDOF models consider structures, which encounter a value-based uprooting of the rooftop in respect to the base. For this situation, the translational mass is basically the concentrated weight partitioned by the speeding up of gravity (32.2 ft/sec2 or 386.4 in./sec2). In any case, cases do emerge in which the rotational movement of the framework is noteworthy. A case of this may be the rotational movement of a rooftop section which has unsymmetrical horizontal backings. Newton's second law of movement expresses that the time rate of progress of the precise force (snapshot of energy) equivalents the torque.

RESPONSE SPECTRA

Reaction spectra are bends plotted between greatest reaction of SDOF framework subjected to determined tremor ground movement and its day and age (or recurrence). Reaction range can be translated as the locus of most extreme reaction of a SDOF framework for given damping proportion. Reaction spectra hence helps in acquiring the pinnacle auxiliary reactions under direct range, which can be utilized for getting sidelong powers created in structure because of seismic tremor along these lines encourages in quake safe plan of structures. Typically, reaction of a SDOF framework is dictated by time space or recurrence area investigation, and for a given day and age of framework, most extreme reaction is picked. This procedure is proceeded for all scope of conceivable eras of SDOF framework. Last plot with framework day and age on x-hub and reaction amount on y-hub is the required reaction spectra. Relating to determined damping proportion and info ground movement. Same process is completed with various damping proportions to acquire general reaction spectra.

The three spectra i.e. uprooting, pseudo speed and pseudo increasing speed give a similar data on the basic reaction. Be that as it may, every one of them gives a physically important amount (allude conditions (6.11) and (6.12)) and consequently, each of the three spectra are valuable in understanding the way of a quake and its impact on the outline. A joined plot demonstrating each of the three of the unearthly quant it is conceivable in view of the relationship that exists between these three amounts. Taking the log of conditions (6.8) and (6.9).

Factor Influencing Response Spectra

The response spectral values depends upon the following parameters,

- Energy release mechanism
- Epicentral distance
- Focal depth
- Soil condition
- Richter magnitude
- Damping in the system
- Time period of the system

LINEAR SEISMIC INVERSION

Backwards displaying is a scientific strategy where the goal is to decide the physical properties of the subsurface of an earth area that has created a given seismogram. Cooke and Schneider (1983) characterized it as figuring of the world's structure and physical parameters from some arrangement of watched seismic information. The hidden supposition in this strategy is that the gathered seismic information are from an earth structure that matches the cross-area processed from the reversal calculation. Some basic earth properties that are reversed for incorporate acoustic speed, arrangement and liquid densities, acoustic impedance, Poisson's proportion, development compressibility, shear unbending nature, porosity, and liquid immersion.

The technique has for quite some time been valuable for geophysicists and can be sorted into two expansive types: Deterministic and stochastic reversal. Deterministic reversal strategies depend on examination of the yield from an earth demonstrate with the watched field information and consistently refreshing the earth display parameters to limit a capacity, which is normally some type of distinction between model yield and field perception. Thusly, this strategy for reversal to which direct reversal falls under is acted like a minimization issue and the acknowledged earth model is the arrangement of model parameters that limits the target work in delivering a numerical seismogram which best contrasts and gathered field seismic information.

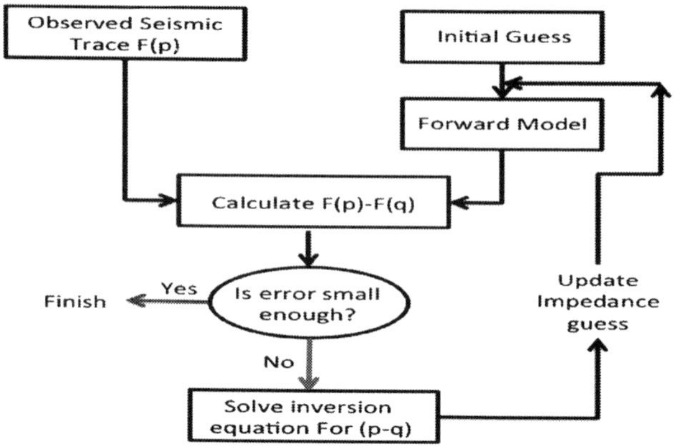

Figure 6.3: Seismic Inversion flowchart

Then again, stochastic reversal techniques are utilized to create compelled models as utilized as a part of supply stream recreation, utilizing geostatistical instruments like kriging. Rather than deterministic reversal strategies which create a solitary arrangement of model parameters, stochastic techniques produce a suite of interchange earth show parameters which all comply with the model limitation. Be that as it may, the two strategies are connected as the aftereffects of deterministic models is the normal of all the conceivable non-remarkable arrangements of stochastic techniques. Since seismic direct reversal is a deterministic reversal strategy, the stochastic technique won't be examined past this point.

Linear inversion

The deterministic way of straight reversal requires a utilitarian relationship which models, as far as the earth show parameters, the seismic variable to be altered. This utilitarian relationship is some numerical model gotten from the crucial laws of material science and is all the more frequently called a forward model. The point of the system is to limit a capacity which is subject to the distinction between the convolution of the forward model with a source wavelet and the field gathered seismic follow. As in the field of improvement, this capacity to be limited is known as the target work and in convectional converse displaying, is just the distinction between the convolved forward model and the seismic follow. As prior specified, diverse sorts of factors can be modified for however for clearness, these factors will be alluded to as the impedance arrangement of the earth demonstrate. In the accompanying subsections, we will depict in more detail, with regards to direct reversal as a minimization issue, the distinctive segments that are important to upset seismic information.

Forward model

The centerpiece of seismic straight reversal is the forward model, which models the era of the exploratory information gathered. As indicated by Wiggins (1972), it gives a useful (computational) connection between the model parameters and figured qualities for the watched follows. Contingent upon the seismic information gathered, this model may fluctuate from the traditional wave conditions for foreseeing molecule relocation or liquid weight for sound wave engendering through shake or liquids, to a few variations of these established conditions. For instance, the forward model in Tarantola (1984)is the wave condition for weight variety in a fluid media amid seismic wave spread while by expecting consistent speed layers with plane interfaces, Kanasewich and Chiu (1985)used the brachistotrone model of John Bernoulli for travel time of a beam along a way. In Cooke and Schneider (1983), the model is a manufactured follow era calculation communicated as in Eqn. 3, where R(t) is created in the Z-area by recursive equation. In whatever shape the forward model shows up, it is imperative that it predicts the gathered field information, as well as models how the information is created. Thus, the forward model by Cooke and Schneider (1983) can only be used to invert CMP data since the model invariably assumes no spreading loss by mimicking the response of a laterally homogeneous earth to a plane-wave source(Sucuoğlu & Akkar, 2014).

1. $$t = \sum_{i=1}^{n} \frac{\left[(x_i - x_{i-1})^2 + (y_i - y_{i-1})^2 + (z_i - z_{i-1})^2\right]^{\frac{1}{2}}}{v_i}$$

where t is ray travel time, x, y, z are depth coordinates and vi is the constant velocity between interfaces i − 1 and i.

2. $$\left[\frac{1}{K(\vec{r})}\frac{\partial^2}{\partial t^2} - \nabla \cdot \left(\frac{1}{\rho(\vec{r})}\nabla\right)\right] U(\vec{r},t) = s(\vec{r},t)$$

where $K(\vec{r})$ represent bulk modulus, $\rho(\vec{r})$ density, $s(\vec{r},t)$ the source of acoustic waves, and $U(\vec{r},t)$ the pressure variation.

3. $s(t) = w(t) * R(t)$

where s(*t*) = synthetic trace, *w*(*t*) = source wavelet, and *R*(*t*) = reflectivity function.

where s(t) = synthetic trace, w(t) = source wavelet, and R(t) = reflectivity function.

Objective function

An essential numerical process in opposite displaying is to limit the goal work, which is a capacity characterized as far as the distinction between the gathered field seismic information and the numerically figured seismic information. Traditional target capacities incorporate the entirety of squared deviations amongst trial and numerical information, as at all squares strategies, the aggregate of the size of the contrast amongst field and numerical information, or some variation of these definitions. Regardless of the definition utilized, numerical arrangement of the reverse issue is gotten as earth model that limit the goal work.

Notwithstanding the goal work, different imperatives like known model parameters and known layer interfaces in a few districts of the earth are likewise joined in the backwards displaying methodology. These requirements, as indicated by Francis 2006, help to decrease non-uniqueness of the reversal arrangement by giving from the earlier data that is not contained in the transformed information while Cooke and Schneider (1983) reports their valuable in controlling commotion and when working in a geophysically surely understood range.

Mathematical analysis of generalized linear inversion procedure

The goal of numerical examination of backwards displaying is to cast the summed up straight reverse issue into a basic framework polynomial math by considering every one of the parts depicted in past areas. viz; forward model, target work and so forth. By and large, the numerically produced seismic information are non-direct elements of the earth display parameters. To expel the non-linearity and make a stage for utilization of direct polynomial math ideas, the forward model is linearized by extension utilizing a Taylor arrangement as done underneath. For more subtle elements see Wiggins (1972),Cooke and Schneider (1983).

Parameterization of the earth model space

Independent of the variable to be transformed for, the world's impedance is a consistent capacity of profundity (or time in seismic information) and for numerical straight reversal procedure to be pertinent for this ceaseless physical model, the persistent properties must be discretized as well as inspected at discrete interims along the profundity of the earth show. Along these lines, the aggregate profundity over which display properties are to be resolved is an essential beginning stage for the discretization. Generally, these properties are tested at close discrete interims over this profundity to guarantee high determination of impedance variety along the world's profundity. The impedance values modified from the calculation speaks to the normal incentive in the discrete interim.

Considering that converse displaying issue is just hypothetically reasonable when the quantity of discrete interims for examining the properties is equivalent to the quantity of perception in the follow to be transformed, a high-determination inspecting will prompt a substantial lattice which will be extremely costly to reverse. Moreover, the framework might be solitary forward conditions, the reversal can be shaky within the sight of clamor and the framework might be under-obliged if parameters other than the essential factors transformed for, are sought. In connection to parameters sought, other than impedance, Cooke and Schneider (1983) offers them to incorporate source wavelet and scale calculate.

At last, by regarding limitations as known impedance values in a few layers or discrete interims, the quantity of obscure impedance qualities to be settled for are decreased, prompting more prominent exactness in the aftereffects of the reversal calculation.

RESPONSE SPECTRA OF EL-CENTRO-1940 EARTHQUAKE GROUND MOTION

The response spectra of the El-Centro, 1940 earthquake ground motion are computed using the exact method described in the earlier Chapter (refer Appendix-I, for digitized values of the earthquake). The spectra are plotted for the three damping ratios i.e. ξ=0.02, 0.05 and 0.1. The displacement, velocity and acceleration spectra are shown in the Figures 6.2, 6.3 and 6.4, respectively.

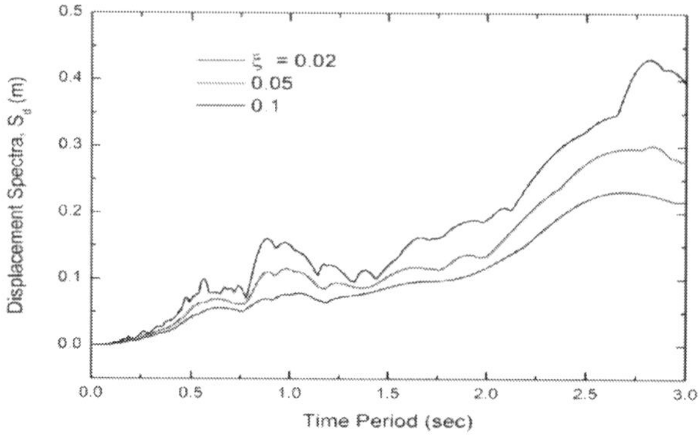

Figure 6.4: Displacement response spectra of El-Centro, 1940 earthquake ground motion.

Further, comparison of the real and pseudo spectra for velocity and acceleration response is shown in the Figure 6.5. As expected, there is no difference between real and pseudo absolute acceleration response spectra. However, the velocity response spectra may have some difference. The digitized values of the response spectra Sd, Sv and Sa of the El-Centro, 1940 earthquake is given in the Appendix – II at an interval of 0.01 sec time period for damping ratio of 2% and 5%.

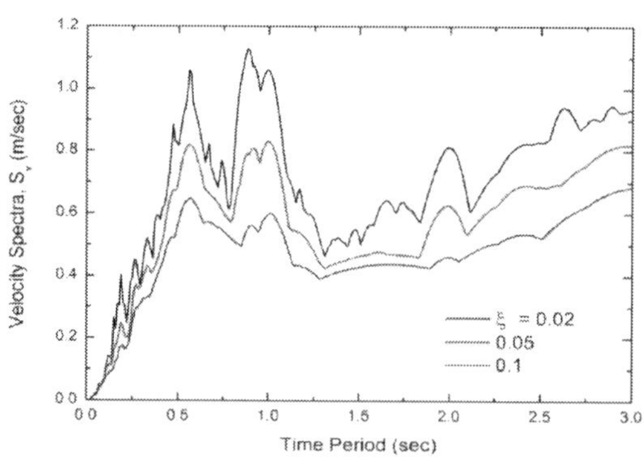

Figure 6.4: Velocity response spectra of El-Centro, 1940 earthquake ground motion.

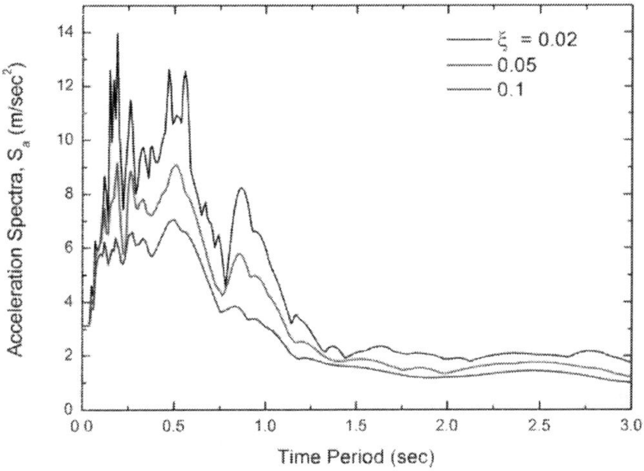

Figure 6.5: Acceleration response spectra of El-Centro, 1940 earthquake ground motion.

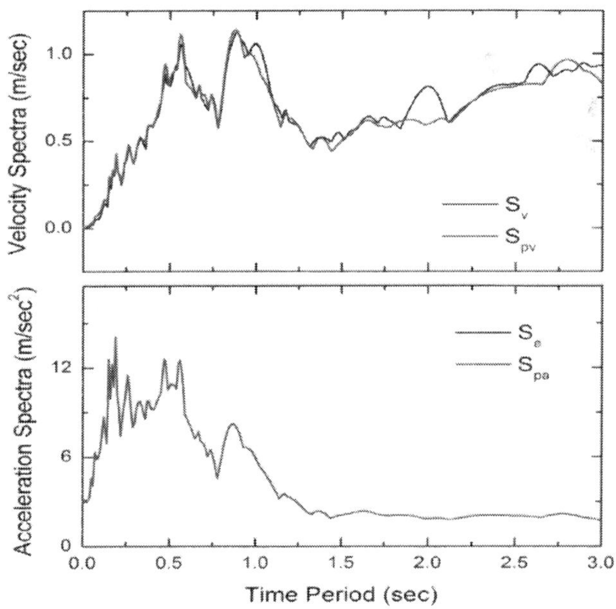

Figure 6.6: Comparison of real and pseudo velocity and acceleration response spectra of El-Centro, 1940 earthquake ground motion (damping ratio=0.02).

RESPONSE SPECTRA METHOD FOR MDOF SYSTEM

Response Analysis of MDOF System

Multi-level of flexibility (MDOF) frameworks are generally dissected utilizing Modal Analysis. A normal MDOF framework with "n" level of opportunity is appeared in Figure (6.8). This framework when subjected to ground movement experiences misshapenness in number of conceivable ways. These twisted shapes are known as methods of vibration or mode shapes. Each shape is vibrating with a specific characteristic recurrence. Add up to one of a kind modes for each MDOF framework are equivalent to the conceivable level of opportunity of framework. The equations of motion for MDOF system is given by

[m]{x (t)} + [c]{x (t)}+ [k]{x (t)} = , [m]{r} x_g (t) (6-12)

where, [m] = Mass matrix (n × n); [k] = Stiffness matix (n × n); [c] = Damping matrix (n × n); {r} = Influence coefficient vector (n×1); { x (t)}= relative displacement vector; { x (t)}= relative velocity vector, { x (t)}= relative acceleration vector, and x g (t) = earthquake ground acceleration.

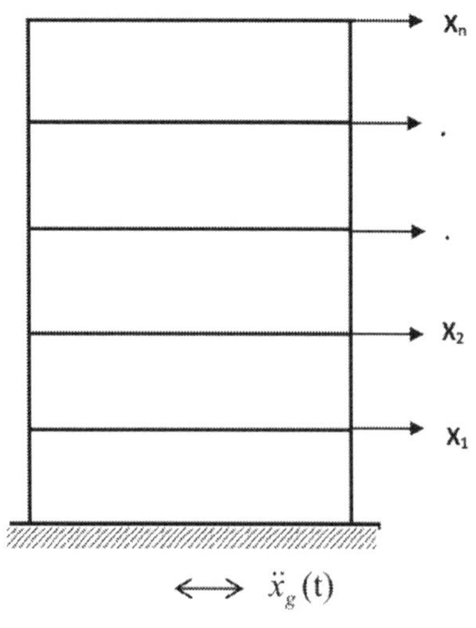

Figure 6.7: MDOF

The required response quantity of interest, ri i.e. (displacement, shear force, bending moment etc.) of the structure can be obtained in each mode of vibration using the maximum response obtained in equations . However, the final maximum response, rmax shall be obtained by combining the response in each mode of vibration using the modal combinations rules. Some of the modal combinations rules commonly used are described here.

Modal Combination Rules

The commonly used methods for obtaining the peak response quantity of interest for a MDOF system are as follows:

- Absolute Sum (ABSSUM) Method,
- Square root of sum of squares (SRSS) method, and
- Complete quadratic combination (CQC) method

In the ABSSUM technique, the pinnacle reactions of the considerable number of modes are included mathematically, expecting that every single modular pinnacle happen at same time. The most extreme reaction is given by. The ABSSUM strategy gives a much preservationist gauge of coming about reaction amount and subsequently gives an upper bound to crest estimation of aggregate reaction. (Chopra, 2007)

DESIGN OF EARTHQUAKE RESISTANT STRUCTURE BASED ON CODAL PROVISIONS

General standards outline reasoning for a plan of seismic tremor-safe structures are the following:

a) The qualities of seismic ground vibrations at any area relies on the greatness of earth tremor, its profundity of center, separation from epicenter, normal for the way through which the waves travel, and the dirt strata on which the structure stands. Ground movements are transcendent in flat bearing.

b) Earthquake created vertical strengths, if critical, as in substantial traverses where differential settlement is not permitted, must be considered.

c) The reaction of a structure to the ground movements is a component of the way of establishment soil, materials size and method of development of structures, and the term and normal for ground movement.

d) The configuration approach is to guarantee that structures have no less than a base quality to withstand minor tremor (DBE), which happen regularly, without harm; oppose direct quake without noteworthy harm however some nonstructural harm may happen, and points that structures withstand real seismic tremor (MCE) without crumple. Genuine powers that showed up on structures are substantially more prominent then the outline powers determined here, however pliability, emerging because of inelastic material conduct and enumerating, and over quality, emerging from the extra save quality in structures well beyond the plan quality are depended upon to represent this distinction in real and plan horizontal strengths.

e) Focus on individuals intended to guarantee that disappointment because of shear or bond does not happen, according to IS:456 and IS:1343.

f) In steel structures, individuals and their associations ought to be proportioned to the point that high flexibility is acquired.

g) The soil structure communication alludes to the impact of the supporting establishment medium on the movement of structure. The structure association may not be considered in the seismic investigation for structures supporting on the stones.

h) The plan parallel powers might be considered in two orthogonal flat headings of the structures. For structures, which have sidelong drive opposing components in two orthogonal bearings just, outline parallel compel must be considered in one course at any given moment. Structures having parallel opposing components in two bearings other than orthogonal should be investigated by statement 2.3.2 IS 1893 (section 1): 2002. Where both even and vertical strengths are considered, stack blends must accord to condition 2.3.3 IS 1893 (section 1): 2002.

i) When an adjustment in inhabitance brings about a structure being renamed to a higher significance calculate (I), the structure should be affirmed to the seismic prerequisites of the new structure with high significance consider.

STRUCTURAL DYNAMICS

Basic examination is, for the most part, concerned with discovering the conduct of a physical structure when subjected to constrain. This activity can include a large load because of the heaviness of things, for example, individuals, furniture, wind, snow, and so forth or some other sort of excitation, for example, a tremor, shaking of the ground because of an impact close-by, and so forth. Fundamentally every one of these heaps are dynamic, including the self-weight of the structure on the grounds that sooner or later in time these heaps were not there. The refinement is made between the dynamic and the static examination on the premise of whether the connected activity has enough increasing speed in contrast with the structure's characteristic recurrence. In the event that a heap is connected adequately gradually, the latency strengths (Newton's first law of movement) can be overlooked and the investigation can be improved as static examination. Auxiliary progression, along these lines, is a kind of basic investigation which covers the conduct of structures subjected to dynamic (activities having high increasing speed) stacking. Dynamic burdens incorporate individuals, wind, waves, movement, tremors, and impacts. Any structure can be subjected to dynamic stacking. Dynamic examination can be utilized to discover dynamic relocations, time history, and modular investigation.

A dynamic examination is likewise identified with the dormancy strengths created by a structure when it is energized by methods for dynamic burdens connected all of a sudden (e.g., wind impacts, blast, and seismic tremor).

A static load is one which differs gradually. A dynamic load is one which changes with time decently fast in contrast with the structure's common recurrence. On the off chance that it changes gradually, the structure's reaction might be resolved with static examination, however in the event that it shifts rapidly (in respect to the structure's capacity to react), the reaction must be resolved with a dynamic investigation.

Dynamic investigation for straightforward structures can be completed physically, yet for complex structures limited component examination can be utilized to compute the mode shapes and frequencies.

Displacements

A dynamic load can have a significantly larger effect than a static load of the same magnitude due to the structure's inability to respond quickly to the loading (by deflecting). The increase in the effect of a dynamic load is given by the dynamic amplification factor (DAF) or dynamic load factor(DLF):

$$DAF = DLF = \frac{u_{\max}}{u_{static}} \qquad (6\text{-}13)$$

where u is the deflection of the structure due to the applied load.

Graphs of dynamic amplification factors vs non-dimensional rise time (tr/T) exist for standard loading functions (for an explanation of rise time, see time history analysis below). Hence the DAF for a given loading can be read from the graph, the static deflection can be easily calculated for simple structures and the dynamic deflection found.

Modal analysis

A modular examination ascertains the recurrence modes or characteristic frequencies of a given framework, yet not really its full-time history reaction to a given information. The common recurrence of a framework is needy just on the firmness of the structure and the mass which takes part with the structure (counting self-weight). It is not reliant on the heap work.

It is helpful to know the modular frequencies of a structure as it enables you to guarantee that the recurrence of any connected occasional stacking won't match with a modular recurrence and thus cause reverberation, which prompts extensive motions.

The technique is:

- Find the regular modes (the shape received by a structure) and normal frequencies
- Calculate the reaction of every mode
- Optionally superpose the reaction of every mode to locate the full modular reaction to a given stacking

Energy method

It is conceivable to compute the recurrence of various mode state of framework physically by the vitality technique. For a given mode state of a different level of opportunity framework you can locate an "equal" mass, firmness and connected compel for a solitary level of flexibility framework. For basic structures the essential mode shapes can be found by investigation, however it is not a preservationist technique. Rayleigh's principle states:

"The frequency ω of an arbitrary mode of vibration, calculated by the energy method, is always greater than, or equal to, the fundamental frequency ωn."

$$\text{Equivalent mass, } M_{eq} = \int M\bar{u}^2 du$$

$$\text{Equivalent stiffness, } k_{eq} = \int EI\left(\frac{d^2\bar{u}}{dx^2}\right)^2 dx$$

$$\text{Equivalent force, } F_{eq} = \int F\bar{u}dx$$

then, as above:

$$\omega = \sqrt{\frac{k_{eq}}{M_{eq}}}$$

Dynamic activities are created on structures by both wind and quakes. In any case, plan for wind strengths and for quake impacts are particularly extraordinary. The natural reasoning of basic plan utilizes constrain as the premise, which is predictable in wind outline, wherein the building is subjected to a weight on its uncovered surface range; this is compel sort stacking. In any case, in tremor plan, the building is subjected to arbitrary movement of the ground at its base , which incites idleness constrains in the building that thusly cause focuses on; this is relocation sort stacking. Another method for communicating this distinction is through the heap twisting bend of the building – the request on the building is compel (i.e., vertical hub) in drive sort stacking forced by wind weight, and dislodging (i.e., level pivot) in uprooting sort stacking forced by seismic tremor shaking.

Twist constrains on the building has a non-zero mean part superposed with a moderately small wavering segment (Figure 5.2). In this way, under wind drives, the building may encounter small changes in the anxiety field, yet inversion of stresses happens just when the bearing of wind turns around, which happens just over a substantial span of time. Then again, the movement of the ground amid the tremor is cyclic about the unbiased position of the structure.

Thus,

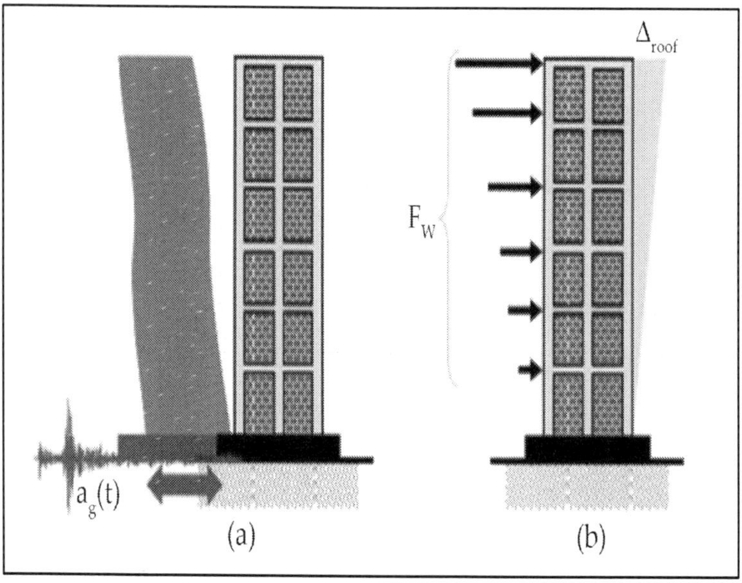

Figure (6.7): Earthquake Ground Movement

The stresses in the building due to seismic actions undergo many complete reversals and that took over the small duration of earthquake.

INERTIAL FORCES IN A STRUCTURE

The seismic body and surface waves make inertial strengths inside the building. Inertial strengths are made inside a question when an outside constrain tries to make it move on the off chance that it is very still or alters its rate or course of movement in the event that it is moving. Inertial drive takes us back to secondary school material science and to Newton's Second Law of Motion, for when a building shakes it is liable to inertial strengths and must comply with this law similarly as though it were a plane, a ship, or a competitor. Newton's Second Law of Motion expresses that an inertial compel, F, squares with mass, M, duplicated by the speeding up. Mass can be accepted as identical (at ground level) to the heaviness of the building, thus this piece of the law clarifies why light structures, for example, wood outline houses, have a tendency to perform preferred in seismic tremors over huge substantial ones, the strengths on the building are less. The speeding up, or the rate of progress of the speed of the waves getting the working under way, decides the rate of the building mass or weight that must be managed as an even compel. An earthquake causes shaking of the ground. So, a building laying on it will encounter movement at its base. From Newton's first law of movement, despite the fact that the base of the building moves with the ground, the rooftop tends to remain in its unique position. However, since the dividers and segments are associated with it, they drag the rooftop alongside them. At the point when the ground movement all of a sudden moves the base of the working to one side, the building moves to the correct in respect to the base , as though it was being pushed to one side by an inconspicuous compel which we call "Idleness Force". Really, there is no push at everything except, due to its mass, the building opposes any movement. All aspects of the building needs to manage inactivity powers. The procedure is substantially more mind boggling than the basic clarification above in light of the fact that the ground moves all the while in two flat and vertical bearings amid a tremor.

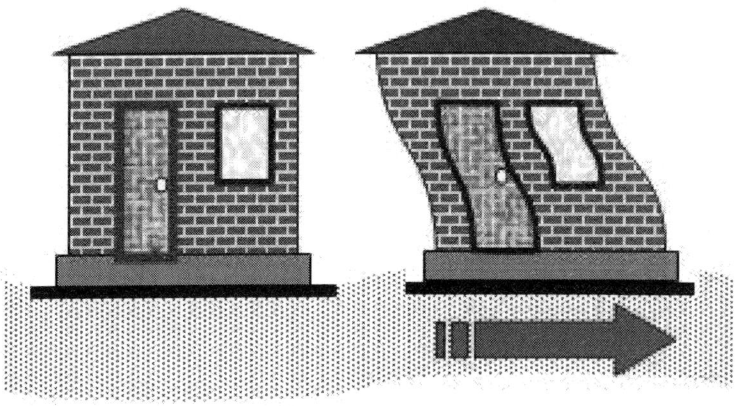

Figure (6.8): Quake Causes shaking of Ground

This is much similar to the circumstance that you are confronted with when the transport you are remaining in all of a sudden begins, your feet move with the transport, however your abdominal area tends to remain back, making you fall in reverse.

This propensity to keep on remaining in the past position is known as dormancy. In the working since the dividers or segments are adaptable, the movement of rooftop is unique in relation to that of ground.

Consider a building, whose rooftop is upheld on sections. Returning to the similarity of yourself on the transport; when the transport all of a sudden begins, you are tossed in reverse as though somebody has connected a compel on the abdominal area. Also, when the ground moves, even the building is tossed in reverse, and the rooftop encounters a compel, called dormancy drive. In the event that the rooftop has the mass M and encounters a quickening a, then from Newton's second law of movement, the inactivity compel F1 is mass M times increasing speed an, and its bearing is inverse to that of the speeding up. Inadequately developed structures start to endure harm at around 10 percent g (or 0.1g). In a direct quake, the influxes of vibration may keep going for a few moments, and increasing velocities might be roughly 0.2g. For individuals on the ground or at the base of a building, the sensations will be fundamentally the same as those of the tenants of the plane in turbulence or travelers remaining in the quick moving train: they feel insecure and may need to take hold of something to help them stay standing. Seismic tremors cause extra caution since when the shaking begins, those encountering it don't know whether it will rapidly end or is the start of a harming and unsafe shudder. Short increasing speeds may, for a small amount of a moment, surpass 1.0g. In the Northridge quake in 1994, a recording station in Tarzana, five miles (8 km) from the epicenter, recorded 1.92g.

Flow of Inertia Forces to Foundations

Under flat shaking of the ground, even latency powers are produced at a level of the mass of the structure (generally arranged at the floor levels). These sidelong idleness powers are exchanged by the floor section to the dividers or the segments, to the establishments, lastly to the dirt framework underneath. Thus, each of this auxiliary component (floor sections, dividers, segments, and establishments) and the associations between them must be intended to securely exchange these latency drives through them.

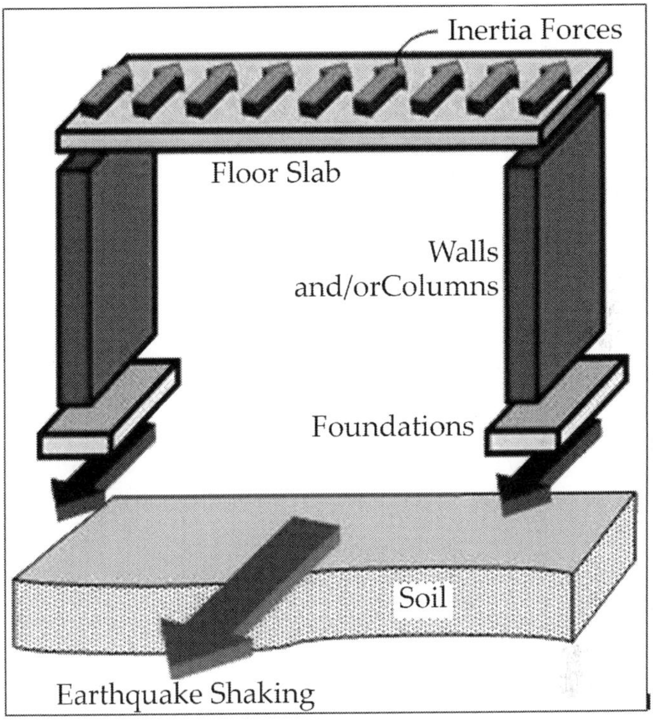

Figure 6.9: Flow of Inertia Forces to Foundations

Dividers or segments are the most basic components in exchanging the inactivity powers. Be that as it may, in customary development, floor chunks and shafts get more care and consideration amid outline and development, than dividers and sections. Dividers are generally thin and frequently made of weak material like brick work.

Internal forces

Oversimplified outline techniques are normally utilized by configuration designers to decide the surmised extent and dissemination of inertial strengths in strengthened solid floor stomachs of multi-story structures. Different scientists have recognized that the usually utilized shortsighted outline technique, the Equivalent Static Analysis (ESA) strategy, now and again, gives a lacking result to the estimation of powers that create in floor stomachs. This examination explores the extent and patterns of powers in solid floor stomachs, with an accentuation on exchange (similarity) powers, under seismic stacking. The initial segment of this exploration examines another pseudo-Equivalent Static Analysis (pESA) technique for deciding inertial powers in floor stomachs. The pESA technique was found to give sufficient outcomes to normal structures in contrast with the more modern non-straight time history investigation comes about. The second portion of this paper looks into that and examines the size of exchange and aggregate strengths inside fortified solid floor stomachs of structures with vertical components of changing firmness (distinctive measured edges or dividers or a mix of dividers and casings). These strengths were examined utilizing non-direct time history investigation. The impacts of: basic malleability, the proportion of the firmness of vertical horizontal drive opposing components, the quality of the association component between the parallel compel opposing frameworks, the auxiliary tallness and different displaying sensitivities on the size of exchange and aggregate strengths were examined. The primary results of this exploration are distinguishing suitable strategies for the plan of floor powers and deciding patterns related with exchange floor powers. Fortified solid stomachs (floors and rooftops) of a structure tie the vertical auxiliary components, (for example, dividers and edges) together to enable structures to oppose outside burdens, for example, gravity and sidelong powers from seismic occasions or wind activity. Floor stomachs assume an essential part of exchanging strengths from the structure to the horizontal constrain opposing components which then exchange the powers from the structure to the ground. The sizes of inside strengths inside solid floor stomachs are significantly more intricate than those expected by some shortsighted strategies utilized in current outline practice, for example, the Equivalent Static Analysis (ESA) strategy. The ESA technique is utilized every now and again in basic outline, because of its straightforwardness and productivity.

Diaphragm forces:

There are two primary sorts of strengths that exist in stomachs, in particular: inertial powers from the increasing velocities of the floors and exchange powers which result from inconsistent disfigurement designs from various parallel constrain opposing frameworks inside the structure.

The sort of parallel drive opposing frameworks and the geometry of the structure will direct which of these powers, inactivity or exchange, will overwhelm. A double or a cross breed structure (blend of dividers and edges) gives sidelong constrain resistance through both edges and auxiliary divider frameworks. A double framework is a good framework because of the capacity to disseminate a lot of vitality.

Two types of internal forces in diaphragms

These are depicted as harmony powers, which are related with remotely connected burdens and similarity strengths related with the distortion of the structure. A look into by Paulay and Priestley has indicated when a double framework is subjected to sidelong loads a type of "battling" exists between both the divider and edge components. This "battling" originates from the differentiating misshapening designs for both edges and dividers.

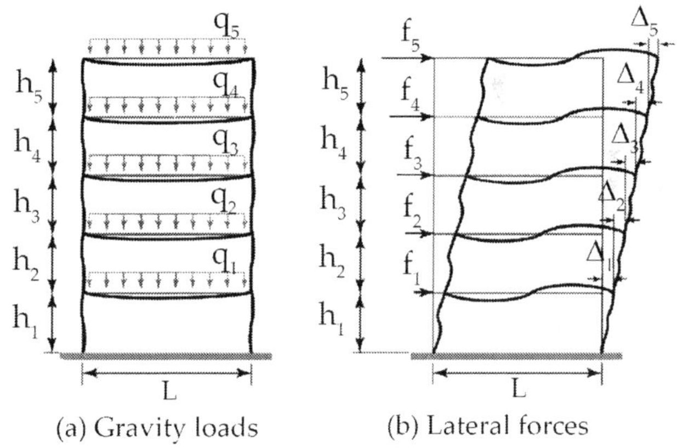

Figure 6.10: Two types of internal forces in diaphragms

The association of the casings and dividers to a stomach requires distortion similarity to exist for the whole structure. This similarity limitation adjusts the general distortion of the structure framing a mix of shear and flexural disfigurement modes which result in the "battling" activity portrayed. This "battling" causes an expansion in powers that are available in the floor stomach. Generally inertial and exchange (similarity) strengths have been dealt with independently. This is not right, both inertial and exchange strengths ought to be considered all the while. As the structure quickens from the inertial constrains it additionally disfigures at the same time. Thusly inertial and exchange strengths ought not be dealt with independently. The issues of size and area of dormancy and exchange stomach strengths are not constrained to "double" frameworks. Tower-to-platform floors and ground floors over storm cellars are further cases. Additionally, structures with a progression of dividers or casings of fluctuating geometries and qualities will have comparative stomach issues.

Transfer forces

These are biggest for casing divider structures where the edge and divider components are of comparable solidness as every component can oppose the misshapenings by the other component. The strengths are observed to be littler when the solidness proportion for the casing divider structure is either little or huge as the more adaptable component gives little resistances to actuate the exchange powers., Inelastic examinations give altogether different outcomes to versatile investigation as far as exchange powers and leftover strengths in the structure. This shows inelastic examinations ought to be completed for these structures., Transfer strengths are observed to be littler for the upper levels of structures where minimal inelastic activity happens., It is found when the inelastic activity is permitted to happen in the association components between the casing and divider (the stomach) the extent of exchange powers and leftover powers lessen., The affectability contemplate showed that it is vital to consider the adaptability of the establishment to precisely decide the floor stomach powers., It was demonstrated that the size of exchange strengths for low seismic districts are unimportant because of the adaptable way of the structure., It is found that in structures with dividers shear disfigurements ought to be represented in the model., It was found the lumped pliancy supposition for divider components prompts wrong outcomes. This suspicion influences the relocations of the divider component which thus influences the improvement of exchange strengths. This exploration has given data on the patterns of powers in floor stomachs and furthermore data on the impacts that some demonstrating suppositions have on the extent of strengths which create inside floor stomachs.

RATE-INDEPENDENT BEHAVIOR OF CLAY

Earth has a crawl conduct; its twisting grows gradually with time under consistent extent of stress. In this manner, the stress–strain bend of earth fluctuates with the rate of stacking. At the point when stacked quicker, dirt uncovers more prominent modulus and higher quality. This perception recommends that mud is a gooey material.

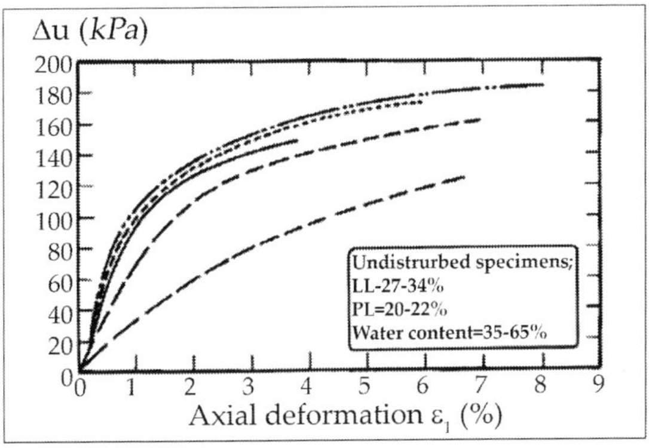

Figure 6.11a: Rate-Independent Behavior of Clay

Then again, the stress–strain–strength qualities of earth is controlled by a grain-to-grain rubbing and synthetic holding and in addition electric association. The extent of these strengths are very little ward of the rate of stacking. Thusly, the stress–strain conduct of mud is half wayrate-subordinate however mostly rate-free.

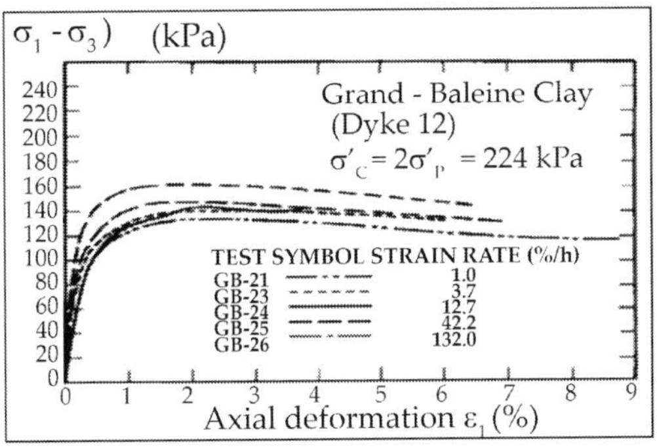

Figure 6.11b: Stress–Strain–Strength Qualities of Earth

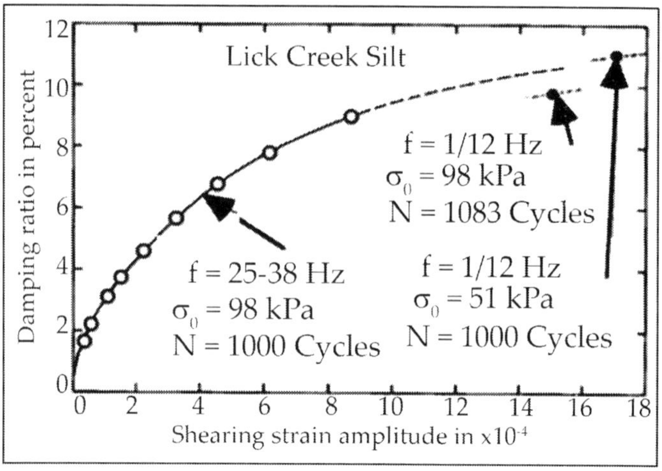

Figure 6.11c: Damping ratio–Strain–Strength Qualities of Earth

Figure 6.10(a,b,c) demonstrates undrained triaxial compression of clay with the axial strain rate varying between 1.0 %/hour and 132 %/hour. Evidently, the faster loading is associated with the greater strength and the less development of excess pore water pressure (the greater effective stress). In spite of the 132 times faster rate of loading, the maximum shear stress (strength) increased by about 25% only. The modulus increased by more or less the same amount as well. Thus, clay behavior is rate dependent to a limited extent. Figure 6.2 illustrates a classic report on the damping ratio of Lick Creek Silt (liquid limit LL =22–34%, plasticity limit PL= 20–27%). Damping ratio stands for the amount of energy loss in cycle due to a hysteretic nature of the loop (Fig. 6.3). A viscous material has a and the rate of loading (frequency in Hz unit). Damping ratio has been known to increase as the amplitude of strain becomes larger. In Fig. 6.2, however, the relationship between damping ratio and the strain amplitude is independent of different frequencies of 25–38 and 1/12 Hz(Bozorgnia & Bertero, 2004).

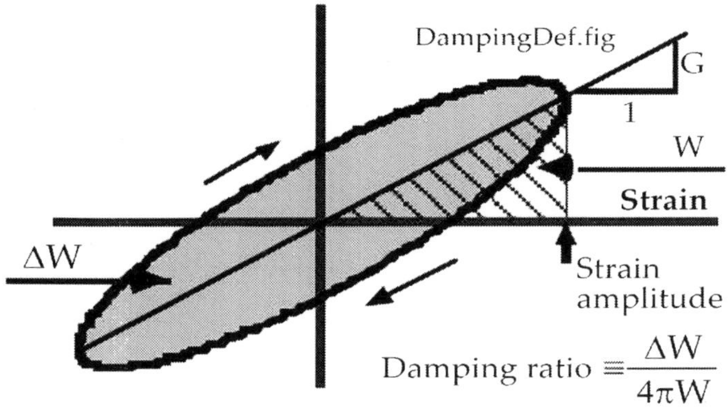

Figure 6.12: Definition of damping ratio in soil dynamics

Significance of Complex Analysis

It might sound odd that numerous, despite the fact that not all, dynamic investigations of ground make utilization of complex numbers: genuine and nonexistent parts. Basic elements barely do this. This practice has a reason in the straightforwardness in examination and a utilization of complex-modulus demonstrating.

An exchange is started with a traditional single-level of-opportunity model with mass, spring, and dashpot, which is known as a Voigt show (Fig. 6.4a). The Voigt model can support a static drive, while the other one in Fig. 6.10b (Maxwell demonstrate) can't. In this manner, the previous model appears to be more like the genuine soil conduct.

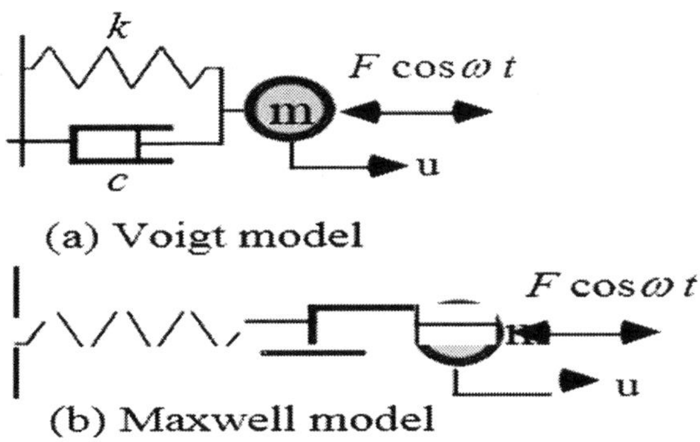

Figure 6.13: Single DOF with real numbers

which is indistinguishable with (6.3). In this way, examinations with genuine and complex numbers are equal when genuine parts of complex numbers are utilized. Be that as it may, the inference of (6.5) is less difficult.

The force–displacement relationship of a Voigt show in this manner determined is delineated in Fig. 6.5. It is apparent that the measure of the circle, which remains for the vitality misfortune per cycle, increments as the stacking recurrence is made higher. Take note of that genuine soil does not show such rate reliance (Fig. 6.2)

The recurrence independency of damping proportion (Sect. 6.1) can likewise be accomplished by changing the dashpot parameter with the stacking recurrence (converse proportionality: 1 recurrence . This thought, be that as it may, is not adequate from a numerical perspective. At the point when the information comprises of consonant movement of various frequencies, the condition of movement is given by

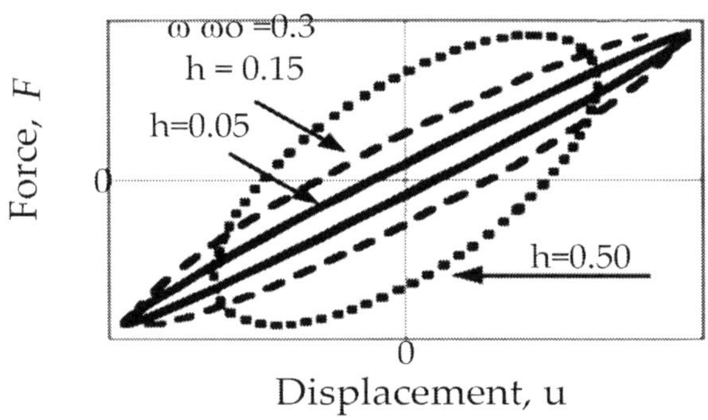

Figure 6.14: Complex Modulus

The act of superposition is supported just when the given differential condition is straight. At the point when the c esteem shifts with h , be that as it may, the condition is not straight anymore and, from the numerical perspectives, the superposition as (6.7) is not sensible any longer. At the end of the day, the seismic tremor reaction is sporadic and its recurrence fluctuates with time in an arbitrary way. Henceforth, it is hard to decide any single number of c parameter. An option technique for examination is either an intricate modulus display or the utilization of nonlinear stress–strain show (e.g., hyperbolic stress–strain demonstrate in or elasto versatility.

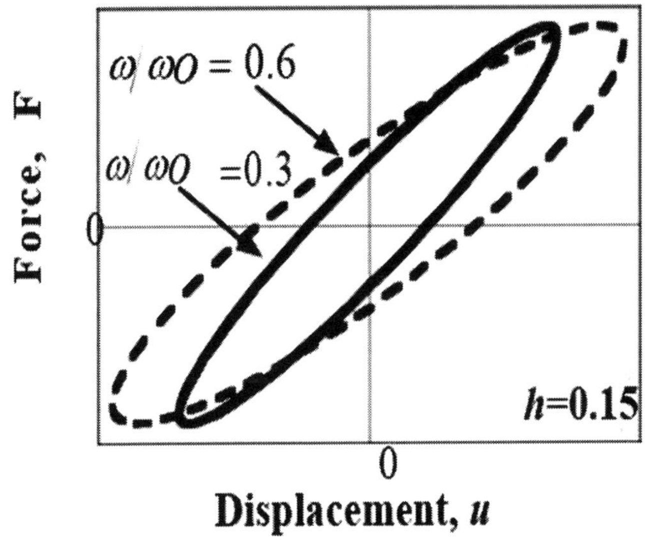

Figure 6.17: Force-Displacement

The vitality dispersal in a perplexing modulus model is represented by the "h" parameter. Figure 6.7 shows that the more noteworthy h esteem makes the circle greater and expands the vitality misfortune per cycle. Moreover, Figure 6.8 uncovers the impacts of stacking recurrence. While the plenteousness of uprooting differs with the recurrence, the damping proportion (Fig. 6.3) is not influenced. This element coordinates the way of soil.

NUMERICAL INTEGRATION OF EQUATION OF MOTION IN THE TIME DOMAIN

A combination of condition of movement in the time space is profitable over the one in the recurrence area in that it can without much of a stretch consider the nonlinearity of material properties and in addition the unpredictable time history of outside load (transient investigation). Then again, being numerical, it can't display the general extent of a concerned dynamic wonder; to be specific, the parameter impacts. Also, nonlinearity of soil properties have been tentatively concentrated, so far, by utilizing steady amplitudes of stress or strain. This circumstance straightforwardly fits an investigation on consonant shaking. In this way, most unique investigation of ground to date utilizes the possibility of consonant shaking. The abnormality of shaking is considered by utilizing the hypothesis of Fourier arrangement (comparable direct technique in Sect. 9.10). It will be conceivable in future that a straightforward and helpful versatility hypothesis is created and turns out to be generally accessible.

This page delineates a basic case of direct combination in the time area. An illustration is taken of a basic condition of movement with a solitary level of-opportunity. A basic investigation is made of a two-layered versatile flat store (Fig. 6.15). The surface layer is proportionate to an alluvium ( ), while the lower layer is a more seasoned store or at times a base shake. In all actuality, there is frequently a solid gravelly layer at the base of alluvium with SPT (N) > 40. The age of two layers is irregular (unconformity) because of topographical History.

The amplitude of ground surface motion is 2E1.

- The free boundary condition at the surface (stress = 0) make the upward (E1) and the downward (F1 = E1) propagations equal to each other in their intensity.
- The amplitude of upward propagation in the base rock is E2. If the base rock is directly exposed to the air (rock outcrop), 2E2 is the surface motion. Hence, 2E2 is called the outcrop motion
- E2 is produced by the earthquake source and the travel path of the earthquake wave. It is the true input motion to the site of concern.
- The baserock motion of E2+F2 has F2 which is the result of wave reflection at z = H and z = 0. F2 is affected by the nature of the surface layer and therefore is not the true INPUT to the site.
- Therefore, E2 has a more essential meaning than E2+F2.

The baserock here may not be a real rock. Being called "the engineering baserock," in reality it is often a pleistocene soil or a soft rock

Interestingly, Amp(E + F) basically thinks about the amplitudes of the surface movement and the movement at the base. The base movement incorporates the descending spread of wave, which is the aftereffect of the reaction of the surface soil, and, hence, Amp(E + F) does not specifically mean the intensification made by neighborhood soil conditions.

At the point when the baserock shows up at the surface, it is known as a stone outcrop. Since a stone mass is thought to be uniform and flexible, the nature of the occurrence movement is indistinguishable at both the outcrop and base shake under alluvium, as appeared by . Figure 6.19 shows the variety of Amp(2E) along the x pivot. Since H fluctuates with x, Amp(2E) alters in the even course also. The most extreme enhancement happens at some separation from the edge of a delicate alluvial plane where the state of reverberation is fulfilled. As indicated by , this most extreme intensification is administered by the impedance proportion. In this way, the gentler store of soil is subjected to the more noteworthy enhancement and the more seismic hazard.

It is normal that a contemplated ground comprises of a wide range of sorts of soils. The surface fill is underlain by clayey and sandy soils potentially together with rock. At the point when this is the situation, it is a typical practice to expect an evenly layered stratification in which layers of various soils are stacked in the level course

Amplification of Motion at the Top of Hill

At the time of 1993 Kushiro-Oki earthquake, the maximum acceleration exceeded 900 Gal (=cm/s2) at the Kushiro Meteorological Observatory (&')*), which was located at the top of a hill (Fig. 6.23). Although this record was surprisingly strong among previous records so-far measured at the ground surface (not inside high buildings), it did not cause structural damage to the observatory building except that shelves and other objects inside fell down to the floor. The reduced damage in the structure was probably because the maximum acceleration continued for a very short time (spiky impact). After this quake, many accelerometers were installed in the Kushiro City and the 1994 Hokkaido-Toho-Oki earthquake

Calculated Amplification Effects of Hilly Topography

The amplification effects of a hill, Amp(2E) and Amp(E+F), are calculated by using the theory as developed in Sect. 6.12. In this section, five types of the hill shape are employed as illustrated in Fig. 6.28. Among them, Amp(2E)=1 and Amp(E+F)= 1 cos(wH Vs) are known for the case of Top width = Base width, B at the far right.

The figured enhancement, Amp(2E), is delineated in. At the point when the top width is shorter than the base width, the intensification is constantly more prominent than solidarity independent of the shaking recurrence. This intensification is equivalent to the adequacy proportion between the slope best and the surface of marsh. It is apparent that the triangular slope with the top width = 0 accomplishes the best enhancement at all the frequencies. Conversely, the top width = B is of Amp(2E)=1 at any recurrence as the hypothesis of uniform flexible medium states. It appears that the supposition of uniform versatility is not all that awful in light of the fact that genuine slopes are made of moderately harder materials than delicate alluvium.

The ascertained substantial intensification for the instance of top width = 0 is sensible. The seismic vitality that is episode at the base ventures upwards. As the width of the slope diminishes at higher heights, the vitality is focused inside a restricted slope width. Since the more noteworthy vitality thickness makes the bigger greatness of shaking , the instance of the zero width at the top shows the biggest intensification. Take note of that Amp(2E) in methodologies the hypothetical estimation of width proportion at higher frequencies. At the point when the top width = B/4 for instance, the width proportion = 1:4 and the shaking vitality thickness = 4:1. Subsequently, the sufficiency proportion is 2:1.

Figure demonstrates Amp(E+F) of a slope geology. It is intriguing that the reverberation in a triangular slope (best width = 0) happens at a recurrence higher than those of more trapezoidal or rectangular geologies.

Observed Topographic Effects

The solid speeding up in the Kushiro seismic tremor movement was seen at the site in Kushiro at the surface and base of the slope (vertical exhibit perception). Notwithstanding Ohta (1995) who demonstrated intensification between the surface and the foot of the slope incline [Amp (2E)], Kanda and Motosaka (1995) and Ishida et al. (1996) indicated expanded speeding up at the surface as contrasted and that in the base layer of the slope. It appears that the intensification between the surface and the base incorporates the topographic enhancement as well as the one because of surface milder material that happens even in a level geology.

The topographic intensification as ascertained in Sect. 6.15 is approved all the more obviously by another arrangement of watched speeding up time histories amid the 1985 Chiba-Toho-Oki tremor. Kurita et al. (2005) recorded speeding up histories in and on a little slope as appeared. The Fourier spectra of quickening records at two stations, No. 3 at the top and No. 5 at the foot, were computed and their proportion was acquired as appeared. It is seen in that that the range proportion that stands for the enhancement of Amp (2E) increments with the shaking recurrence. This finding is reliable with the discoveries. The aggregate fall of houses on a patio slope in Balakot of Pakistan amid the 2005 North Pakistan tremor might be the aftereffects of this topographic enhancement.

Figure 6.18: Completely destroyed houses on terrace hill in Balakot of Pakistan during 2005 earthquake

7

CHAPTER

INNOVATIVE STRATEGIES IN EARTHQUAKE ENGINEERING

INTRODUCTION

The section includes initial comments in regards to the significance of creating and actualizing inventive procedures in seismic tremor designing.

The principle motivations behind the over two segments are (1) to present "creative systems," the purposes behind their need and the primary promising procedures and methods that have been proposed and utilized and (2) to talk about in detail the advancement and uses of one gathering of methodologies and strategies in Japan.

Point by point exchanges of advancements and utilizations of seismic segregation systems are exhibited and those of vitality dissemination techniques and gadgets are introduced in Chapter 12. In this way, examines the advancement and utilization of vitality dispersal systems in Japan that have been founded on the idea of "harm controlled structures" (DCS) with specific consideration on the utilization of the method or component of "unbonded supports."

INTRODUCTORY REMARKS REGARDING IMPORTANCE OF DEVELOPING AND IMPLEMENTING INNOVATIVE STRATEGIES IN EARTHQUAKE ENGINEERING

Main Objectives

The fundamental targets of this segment are to:

- Make a few remarks about
- The part and significance of quake building in controlling the seismic dangers in our urban and rustic ranges
- Possible purposes behind watched increment in seismic dangers
- The improvement of a reasonable structure for execution based seismic building (P-BSE)

Examine quickly the methodologies and approaches for P-BSD with specific accentuation on

- The disadvantages of the current regular U.S. configuration code approaches
- The P-BSD approach by Wada et al. (1992) and Connor and Wada (1992)
- P-BSE approach in light of the utilization of vitality idea
- inference of vitality conditions
- preferences of utilizing vitality ideas in P-BSE of structures
- the vitality based idea can be utilized as a part of P-BSE
- methodologies for reaction and harm control (or defensive frameworks)

Talk about what inventive methodologies are and:

- offer an order and meanings of various frameworks that have been proposed and utilized
- specify late advances in various imaginative auxiliary frameworks

Role and Importance of Earthquake Engineering in Controlling Seismic Risks

Section 1 recommends that "quake building" can be characterized as the branch of designing that envelops all specialized and nontechnical endeavors to diminish the seismic hazard in urban and rustic zones to levels that are socially and financially satisfactory to the group in such territories. It has likewise called attention to that we are a long way from decreasing the seismic dangers in our manufactured condition to socio-monetarily satisfactory levels. Quakes are inescapable, yet seismic tremor debacles can be controlled. The blame cracks from which huge quakes start don't straightforwardly slaughter individuals or even the burst without anyone else's input does not bring about extraordinary financial misfortunes. The reason for a large portion of the wounds and monetary misfortunes is the collaboration of the seismic tremor ground movements (EQGMs) with the manufactured condition. What is required is the lessening of seismic dangers by controlling the constructed condition and this ought to be a definitive objective of quake building(Huang, Bertero, & Wada, 2004).

Observed Increase in Seismic Risks

In a few regions of the world, despite the fact that seismicity stays steady, the fiasco capability of a tremor, and in this way the seismic hazard, increments. An unmistakable case is an expansion in monetary misfortunes in California accordingly of late seismic occasions. For instance, look at what occurred amid the 1989 Loma Prieta quake with what occurred as an outcome of the colossal 1906 San Francisco seismic tremor, and in addition what occurred in the 1994 Northridge quake when contrasted and what occurred as a result of the 1971 San Fernando seismic tremor. In spite of the fact that the field examinations of the outcome of the above seismic tremors in California have plainly exhibited that the greater part of the offices that have been outlined and built by the most recent construction regulation arrangements could give life assurance, the monetary misfortunes because of the harm to basic and nonstructural segments, building substance, life savers, cost of repair or recovery or both, and misfortune because of downtime and business intrusions were extensively higher than anticipated. Comparable conclusions have been come to from investigation of what occurred in late seismic tremors in Japan (Kobe 1995), Turkey (Kocaeli and Düzce 1999) and Taiwan (Chi-Chi 1999).

In spite of the generally gentle level of ground shaking created, the 1989 M6.9 Loma Prieta tremor brought about more than $ 8 billion in direct harm, with a few scaffolds and viaducts enduring genuine devastation (incomplete or add up to breakdown). Albeit no death toll occurred in present day structures, the financial misfortune was judged by both the basic building experts and open arrangement producers, as being too substantial for this direct occasion. A need was recognized for new building, scaffold and viaduct plan and development methods, which could better meet society's prerequisite that property and business interference misfortunes in direct quakes ought to be controlled to satisfactory levels. The need to grow the extent of methodology used to assess and restore existing structures, extensions and viaducts and in addition other structural designing offices was likewise recognized.

Development of Conceptual Framework for P-BSE

In perspective of the above necessities, the Structural Engineers Association of California (SEAOC), which has been effectively occupied with the advancement of the seismic outline and development arrangements in the U.S. construction standards, especially in the Uniform Building Code, perceived the requirement for the advancement of another execution based approach for the seismic plan and development of structures. Along these lines, in 1992, the SEAOC directorate built up the Vision 2000 Committee, to build up a structure for an up and coming era of execution based seismic code. Be that as it may, very little advance was made till 1994.

On January 17, 1994, the M6.7 Northridge seismic tremor occurred, bringing about misfortunes evaluated at around $20 to 30 billion; misfortunes more extreme than those subsequent from the Loma Prieta quake. Designers and open approach creators again verified that it is unsatisfactory to experience this greatness of misfortune in these generally regular and direct occasions. Confronted with a need to repair and recreate a huge number of structures, the California Office of Emergency Services contracted SEAOC to create proposals for execution based outline and development methodology that can be utilized as a part of practice sooner rather than later. In 1 year the Vision 2000 Committee created between time execution based building proposals (SEAOC, 1995)(Huang et al., 2004).

As is talked about in points of interest in Chapters 8 and 9, the SEAOC (1995) suggestions contain a theoretical structure for P-BSE. Perceiving that the most squeezing issues around then were the need to create dependable systems for the appraisal of seismic defenselessness of existing structures and for the repair, recovery or overhauling of those judged to be risky, FEMA upheld a venture to create "Rules for Seismic Rehabilitation of Buildings," which brought about the distribution of such rules in 1997 (FEMA-273 and 274, 1997) and in the current Prestandard FEMA-356 (2000).

Approaches and Methodologies for Performance-Based Seismic Design (P-BSD)

This area talks about quickly:

1. P-BSD approach considered in the theoretical structure for P-BSE prescribed via SEAOC (1995) and downsides of the current traditional U.S. configuration approach.
2. The P-BSD approach by Wada et al. (1992) and Connor and Wada (1992).
3. P-BSE approach in light of the utilization of vitality idea with specific accentuation on determination of vitality conditions, focal points of utilizing vitality ideas in P-BSE of structures, how the energy based idea can be utilized as a part of P-BSE and inventive methodologies for reaction and harm control (or defensive frameworks).

P-BSD Approach Considered in the Conceptual Framework for P-BSE Recommended by SEAOC (1995) and Drawbacks of Existing Conventional U.S. Design Approach

As examined in Chapter 8, the P-BSE system, outlined in Figure 7.1, envelops the full scope of building exercises important to make structures with unsurprising seismic execution. Execution based seismic building starts with the determination of execution targets and the recognizable proof of the potential seismic perils to have the capacity to set up the outline quakes, and whether the site chose is reasonable for tremor safe development. This system then incorporates the plan stage, which incorporates the theoretical, preparatory and last numerical outline and the outline worthiness checks and audits. At long last, the structure finishes up with quality affirmation amid development and checking of building support and its inhabitance (or capacity) after development. From this system, plainly the auxiliary plan stage comprises of a grouping of three stages and that before beginning such strides, it is important to choose an appropriate approach that will finish the chose particular outline execution destinations (execution outline criteria). Along these lines, the architect must have a decent general handle of all the principle parts of the seismic tremor issues and of the rationalities, methodologies and procedures accessible practically speaking to effectively lead the aggregate plan prepare.

Basic audits of the accessible outline methods of insight and methodologies are condensed in Chapter 8 and in Bertero (1992a, 1992b, 1997a, 1997b), and are talked about in detail in SEAOC Vision 2000 Committee Report (1995) and furthermore in Oliveto (2002, Chapter 1). From these audits it turns out to be evident that to enhance the real execution, and especially harm control of the genuine built offices, there is a need to change the current U.S. seismic plan code approach, which depends on the outline for the quality limit requested by the predetermined base shear, which thusly relies on upon the outline phantom increasing speed. In this approach is alluded to as "the customary or ordinary plan approach."

Examinations of flexible reaction spectra of the EQGMs recorded amid critical seismic tremors have unmistakably demonstrated that it is not financially attainable to plan and develop standard inhabitance structures that will stay versatile under expected serious EQGMs. Along these lines, to acquire safe and furthermore socially and monetarily adequate developments when they are subjected to serious EQGMs, it is attractive to decrease the requested versatile quality and, henceforth, enabling the structure to experience controllable inelastic (plastic) distortions.

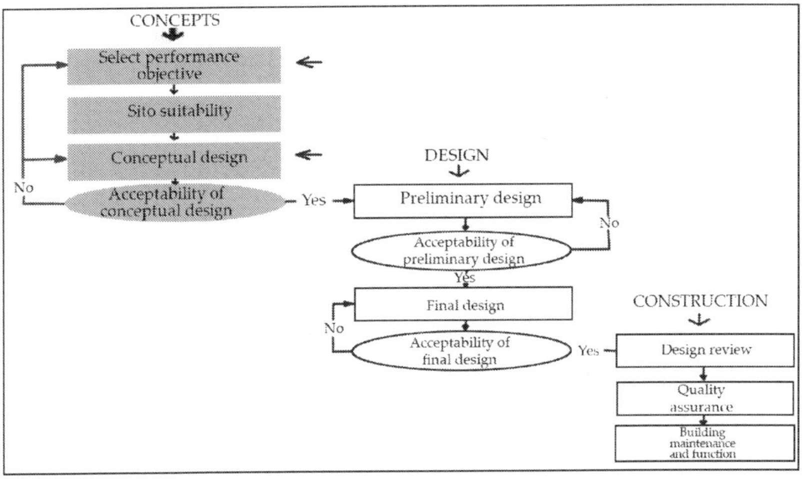

Figure 7.1: Conceptual framework for P-BSE. (Adapted from SEAOC Vision 2000 Committee (1995). Performance-based seismic engineering of buildings

For the structure to grow such controllable plastic disfigurements it is important to give the customary structures adequate durability (vitality scattering limit through yielding) to keep away from an early semi weak disappointment. Such durability by and by is normally spoken to by the supposed "worldwide pliability proportion, μ," which is utilized as one a player in the purported "reaction alteration figure, R," to decrease the direct outline (quality) spectra to decide the required base shear quality.

While there is probably it is alluring to give the structure the most extreme pliability proportion that is financially doable, it must be perceived that physical flexibility by yielding of current customary auxiliary materials prompts harm amid a serious EQGM. The bigger is the malleability the higher is the harm. Therefore, while it is attractive, and ought to likewise be prescribed to furnish the structure with the greatest flexibility proportion that is financially attainable, this high malleability proportion ought not to be abused to exorbitantly lessen the required yielding quality. At the end of the day, it is important to control the requested worldwide pliability by giving adequate worldwide yielding quality limit to such an extent that the structure is furnished with the required sturdiness to achieve a sheltered working as well as to stay away from exorbitant harm. Figure outlines the noteworthy harm that is created when structure and auxiliary parts are subjected to vast malleability requests.

In light of the previously mentioned downsides of the customary code approach, noteworthy endeavors have been dedicated to discovering more solid methodologies and philosophies to control auxiliary seismic execution. In this part just the accompanying general methodologies that have been suggested in Japan and United States are quickly presented.

Performance-Based Seismic Design Approach Proposed by Wada et al. (1992) and Connor and Wada (1992)

In 1992, Wada et al. examined the creative approach of harm tolerant structures (DTS), or damage control structures (DCS), which is hypothetically in light of the execution based outline philosophy proposed by Connor and Wada (1992), Connor and Klink (1996) and of late by Connor (2003). This idea and its innovation, which has been widely connected in tall structures in Japan (more than 60 m (197 ft)) and begun to be connected in outline and development of new structures in the United States, are talked about in this part.

Performance-Based Seismic Design Approach Based on the Use of Energy Concepts

As opposed to the customary or traditional procedure of utilizing the removal pliability $\mu\delta$ to build up an inelastic outline range to decide the base yielding quality for the seismic plan of the structure, an option was proposed by Housner (1956) to utilize a vitality approach. In spite of the fact that assessments have been made of info vitality to single-level of-flexibility (SDOF) frameworks (Berg and Tbomaides, 1960), and even of multi-degrees-of-opportunity (MDOF) frameworks, for example, steel structures planned in the 1960s for a portion of the current recorded EQGMs (Anderson and Bertero, 1969). It is just as of late that this approach has increased broad consideration (Akiyama, 1985). This outline technique depends on the start that the vitality request amid a quake can be anticipated and the vitality supply of an auxiliary component and of basic framework can be set up. In an acceptable outline, the vitality supply is bigger than vitality request.

Figure 7.2: Photos illustrating significant damage to structures and structural components when they have to develop large ductilities

As talked about by Bertero (1992a), to create dependable plan techniques in view of vitality approach, it is important to determine the vitality conditions. Albeit genuine structures are generally MDOFS, to encourage the investigation and comprehension of the physical importance of the vitality approach, it is helpful to first determine the vitality conditions for SDOFS and after that for MDOFS.

Energy Equations

Uang and Bertero (1988) give detailed discussion of the derivation of two basic energy equations (absolute and relative) starting directly from the equilibrium equation of an inelastic viscously damped SDOFS subjected to an EQGM.

The "absolute" energy equation is

$$E_I=E_K+E_S+E_{H\xi}+E_{H\mu} \qquad (7.1)$$

where,

E_I="absolute" input energy E_K="absolute" kinetic energy E_S=elastic strain energy

$E_{H\xi}$=hysteretic dissipated energy due to viscous damping $E_{H\mu}$=hysteretic dissipated energy due to yielding or friction or both

E_I is defined "absolute" input energy because it depends on the absolute ground acceleration. Physically, it represents the inertia force applied to the structure. This force, which is equal to the restoring force plus damping force, is the same as the total force applied to the structure foundation. Therefore, E_I represents the work done by the total base shear at the foundation on the foundation displacement.

The "relative" energy equation is

$$E'_1 = E'_k + E_s + E_{H\zeta} + E_{H\mu} \tag{7.2}$$

where,

E'_I = "relative" input energy

E'_I = "Relative" kinetic energy

E'_I is defined as the "relative" input energy because it represents the work done by the static equivalent external force (which is the product of the mass and ground acceleration) on equivalent fixed-base system, that is, it neglects the effect of the rigid body translation of the structure.

Input Energy to MDOFS

The E_I for an N-story building (N levels above the foundation) can be calculated as follows (Uang and Bertero, 1988)

$$E_1 = \int \left(\sum_{i=1}^{N} m_i \left(a_l \right)_i \right) du_g \tag{7.3}$$

where m_i is the lumped mass associated with the ith floor, $(a_t)_i$ is the total acceleration at the ith floor and u_g is the ground displacement. In other words, E_I is the summation of the work done by the total inertia force $(m\ (a_t)_i)$ at each floor through the ground displacement, u_g. Analysis of results obtained from experiments conducted on medium rise steel dual systems indicates that the EI to a multistory building can be estimated with sufficient practical accuracy by calculating the E_I of SDOF systems using the fundamental period of the multistory structure.

Advantages of Using Energy Concepts in Seismic Design of Structures

Equation 7.1 can be rewritten as

EI=EE+ED (7.4a)

$E_I=E_K+E_S+E_{H\xi}+E_{H\mu}$ (7.4b)

where $E_E=(E_K+E_S)$ can be considered as the stored elastic energy and $E_D=(E_{H\xi}+E_{H\mu})$ the dissipated energy. Comparing this equation with the basic design equation,

Demands≤Supplies

it becomes clear that E_I represents the demands and the summation of E_E+E_D represents the supplies. Equation 7.4b points out to the designer that to obtain an efficient design, the first step is to have a good estimate of E_I for the critical EQGM. The designer must then analyze if it is possible to balance this demand with just the elastic behavior of the structure to be designed or will it be convenient to attempt to dissipate as much as possible some of the $E_{I,}$ i.e., using E_D. As revealed by Equation 7.4b there are three ways of increasing E_D: (1) by increasing the linear viscous damping, $E_{H\xi}$, (2) by increasing the hysteretic energy $E_{H\mu}$ or (3) a combination of increasing $E_{H\xi}$ and

By and by, it is a typical practice to simply attempt to build the EHμ however much as could be expected through inelastic (plastic) conduct of the structure, which suggests harm of the auxiliary individuals. Just since the 1980s has it been perceived that it is conceivable to increment essentially EHμ and control the harm all through the structure using vitality scattering gadgets. Moreover, as examined by Bertero (1992b) (in a dialog of the utilization of a vitality condition for reasonable choice of seismic redesigning systems for existing dangerous structures) expanding ED by expanding EHξ, as opposed to EHμ, has the considerable preferred standpoint that it can control the conduct of the structure under both wellbeing and administration levels of EQGMs.

Assuming in fact or financially (or both) it is impractical to adjust the required EI either through EE alone or EE+ED, the originator has the choice of endeavoring to control (diminish) the EI to the structure. This should be possible by base segregation systems. A mix of controlling (diminishing) the EI by base confinement systems and expanding ED by the utilization of vitality dissemination gadgets is a promising inventive methodology not just to achieve proficient tremor safe outline and quake safe development of new offices, additionally for the seismic overhauling of existing perilous structures (Bertero and Whit taker, 1989).

How Can the Energy Concepts Be Used in Performance-Based Seismic Design (P-BSD)?

A reasonable strategy for the extensive seismic plan approach was produced and connected for outline of structures quite a long while prior (Bertero, 1992b; Bertero and Bertero, 1992). This procedure has been somewhat adjusted for execution based seismic outline (P-BSD) as per the suggested plan target network and the structure for execution based seismic designing (P-BSE) shown in Figure 7.1. The applied philosophy was produced as per the far-reaching configuration approach and the around the world acknowledged seismic

outline logic and depends on the utilization of vitality ideas and major standards of auxiliary elements. It considers from the earliest starting point of the tremor safe outline technique, the concurrent requests (counting torsional impacts) for quality, misshapening and rate of twisting and their consolidated consequences for the EI and on the required and provided vitality limits (as exhibited in Equation 7.4) of the whole office framework. The iterative strategy required in the far-reaching configuration approach proposed for P-BSE is represented in Figure 7.1, and its nitty gritty talks are given in Appendix B of the Vision 2000 report (SEAOC, 1995), and in Chapter 8. As demonstrated in Figure 7.1, the P-BSD is led in three stages beginning with the calculated plan, as characterized next.

Meaning of calculated quake safe outline of structures: Conceptual general tremor safe plan is the shirking or minimization of issues that can be made by the basic mix of the seismic impacts with other likely excitations (stacking conditions), utilizing a physical comprehension of the mechanical (dynamic) conduct of the whole office, instead of numerical calculations (Bertero, 1979, 1982, 1995, 1997a, 2002; Bertero and Bertero, 1992; SEAOC, 1995). Calculated general outline of the office includes not just the decision of general arrangement and size of the office additionally the determination of the auxiliary format, basic framework, basic material, kind of nonstructural segments (especially those that could end up noticeably unexpected basic parts), hardware, utility lines and the establishment framework.

On account of structures, both the modeler and the architect need to see how plan choices with respect to building format, and sum and appropriation of the mass, may have genuine seismic impacts on the structure. The inertial powers rely on upon the mass (sum and conveyance), the damping and the auxiliary qualities (firmness, yielding quality, greatest quality, vitality assimilation and vitality dissemination limits). In spite of the fact that there is no all-inclusive perfect building and auxiliary arrangement, certain essential standards of seismic tremor safe outline can be utilized as rules to choose satisfactory building and basic designs, and also proficient basic establishment sorts and frameworks, nonstructural parts, utility lines and their separate materials. In this lone the accompanying thing is said. Choice of the correct genuine framework and way to deal with be utilized as a part of the preparatory P-BSD and the last measuring and itemizing of the auxiliary individuals ought to be founded on (or possibly considered), the utilization of vitality ideas using the vitality adjust condition, as per Equation 7.4a and Equation 74b. The stream outline exhibited in Figure 7.3 can encourage the audit talk on how the above vitality conditions can be actualized. As pointed out already, the past execution of structures has demonstrated that legitimate use of the customary approach in light of simply the utilization of the ordinary or conventional U.S. code plan criteria could bring about building development that gives "Life-Safety" against extreme EQGMs. Nonetheless, by and large, it is hard to control the harm to socially and financially worthy levels. The primary explanation behind this trouble is that the present code outline criteria depend on decreasing the flexible base shear quality requested by the EQGM. This decrease in quality can be accomplished by requiring that a lot of EI is dispersed through EHμ (i.e., through yielding or plastic disfigurement) that could happen at certain basic locales of the structure, which in this way would endure changeless harm. The key issues with this approach are to begin with, dependable choice of where the basic districts are found and second, to measurement and detail these basic locales so they can scatter adequate EHμ so that the office can be intended for inertial strengths fundamentally lower than those required for accomplishing exclusively flexible reaction. While examining the favorable circumstances and disservices of this approach, it must be remembered to control the EHμ that would be produced in each basic district where the inelastic misshapenings would happen. Pliability through yielding of materials is typically connected with harm; the bigger the physical flexibility, the more prominent will be the harm.

Despite the fact that the use of the limit outline rationality (Park and Paulay, 1975; Paulay and Priestley, 1992) can essentially control the area of the basic districts, there are still a few troubles with its application in profoundly repetitive, and especially unpredictable, structures, and in the expectation of genuine hysteretic conduct at basic areas (Bertero and Bertero, 1992). As represented in Figure 7.3, on account of a minute safe casing structure, the basic areas are generally situated in the shaft close or neighboring the bar section joints. On account of the extensive variety in the genuine mechanical qualities of the basic materials concerning the ostensible qualities indicated in the codes, it is exceptionally hard to find the basic districts and furthermore to decide the length of these areas and to do their legitimate specifying. Besides, regardless of the possibility that the basic districts are legitimately point by point in the drawings, a portion of the subtle elements are extremely hard to do dependably in the field. Hence, it is not amazing to see, in trials, that the hysteretic conduct of the basic areas begin to corrupt rashly and that this marvel increments fundamentally with the reiteration of inversion cycles of inelastic misshapenings. This debasement is extremely hard to foresee due to its high affectability to the quality control of the materials and their genuine itemizing, i.e., workmanship in the field. In this way, it is extremely hard to control the genuine execution (harm) for the whole office. Moreover, to accomplish huge EHμ in the basic areas of a structure more often than not requires huge inter-story float. These extensive inter-story floats result in significant harm to nonstructural components, (for example, infill dividers, parcels, lintels and roofs) and additionally to the gear and substance and the utility lines of the office.

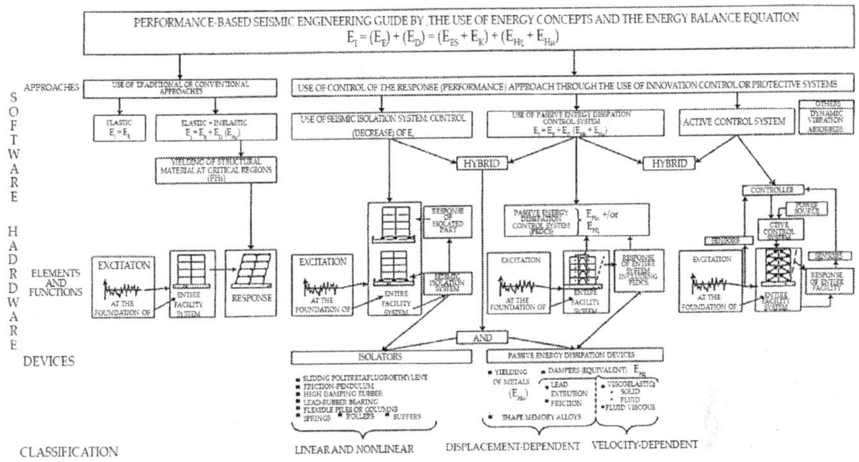

Figure 7.3: Flow chart of the conventional and innovative approaches, strategies and techniques for P-BSD. (From Bertero, V.V., 1997

Assurance of nonstructural parts, gear and building substance. As it has been talked about in a few distributions (Bertero, 1980, 1992a, 1996, 2002; Gillengerten, 2001; Johnson, 2003; Mayes, 1996, Mayes and Naeim, 2001), and furthermore examined in Chapters 8, 9, 11, 12 and 19, a standout amongst the most troublesome issues to address from a customary building-outline perspective is lessening harm to nonstructural parts, gear, utility lines and building substance. This is all the time disregarded and when tended to can be extremely costly to join in traditional plan. There are two essential reaction parameters that cause nonstructural harm. The first is identified with the relative float between floors (as a rule alluded to as inter-story float) and the second is the floor increasing speeds. Together, these two parameters make harm the building substance, compositional exteriors, allotments, funneling and ventilation work, roofs, building gear and lifts. There are two distinctive plan rationalities wrangled inside the basic building calling that arrangement with limiting nonstructural harm. One contends that solid structures decrease inter-story float; however, their rivals express that they deliver high-floor increasing velocities. The other school of thought contends that adaptable structures are the arrangement, since they draw in less compel and have a tendency to diminish floor increasing speeds. In spite of the fact that this is valid, their adversaries express that adaptable structures have considerably higher inter-story floats, and this complements harm to parts that are touchy to float. Plainly what is required is a plan idea that will bring about a decrease of both inter-story float and floor increasing velocities, i.e., new dependable inventive methodologies that will bring about control of the coveted seismic reaction.

Innovative Approaches for Response Control (or Protective Systems)

Recorded surveys of the endeavors to control the seismic reaction and harm of structures, especially their crumple, demonstrate that few such endeavors were made in early human progress (Neddham, 1971). Be that as it may, in the United States the majority of the imaginative methodologies and techniques, in view of the logical standards of tremor designing to control the seismic reaction of genuine structures, have been proposed since the 1980s; few of them have been effectively connected in different nations before (since 1970s).

What Are the Innovative Approaches?

As talked about in Oliveto (2002, Chapters 1 and 2), creative ways to deal with the issues of seismic designing depend on the expansion of extraordinary mechanical or auxiliary gadgets to what has been brought in the United States "customary basic framework" and in Japan called "essential basic framework." The fundamental goal of these methodologies is to enhance the control of the seismic reaction (floor increasing speed, distortions, float and plastic misshapening). This can be accomplished either by controlling the EI to a huge piece of these offices or by limiting (or notwithstanding disposing of) its harming impacts by controlling ED (especially EHμ) request on the customary essential auxiliary individuals or by both of these. By controlling these sorts of vitality request and vitality supplies, it is conceivable to control the inter-story float lists and to bring down the increasing speeds, and hence control the harm to the essential basic individuals and additionally lessen the harm to the nonstructural parts, gear, utility lines and substance of these offices. This control (decrease) of the harm to the whole office framework is the fundamental motivation behind why these imaginative methodologies are generally additionally called defensive methodologies and the framework and the gadgets utilized are typically called defensive frameworks and gadgets.

The above procedure or idea of controlling the dynamic reaction (auxiliary vibrations) by including mechanical or basic gadgets is not new; it has been connected in numerous different branches of designing. As has been called attention to in the relevant writing (Hanson et al., 1993), some of these applications incorporate safeguards for vehicles, gear vibration isolators, pipe limitations and snubbers, stun disengagement gadgets to alleviate impact impacts and mass damping frameworks to control wind-incited vibration in structures. In any case, generally few of these gadgets have been connected particularly to control (decrease) the seismic reactions. Just as of late have there been some huge advances in the refinement of accessible mechanical and auxiliary gadget equipment and in the improvement of new ones for application to the quake safe outline and tremor safe development of new polite designing offices and for seismic redesigning of existing frameworks, especially scaffolds, viaducts and structures(Huang et al., 2004).

Classification and Definition of Innovative Control (or Protective) Systems for Seismic Effects

Classification

As indicated in Figure 7.3, the different innovative systems that are available for control of the seismic response of civil engineering facilities can be grouped under the following classification: seismic isolation, passive energy dissipation, active and hybrid control systems and others. This classification is practically the same as has been adopted by ATC-17–1 (1993) and later by FEMA-273 and -274 (1997) and FEMA356 (2000).

Definitions

ATC-17–1 (1993) has adopted the following system definitions:

- Seismic confinement frameworks decouple building and scaffold structures from the harming segments of EQGMs. Such frameworks require that the structure be bolstered on discrete isolators, which detune the basic framework and include generous damping, cases incorporate elastomeric, contact and sliding gadgets.

- Passive vitality scattering control frameworks "include damping" to structures such that they fundamentally diminish reaction (and harm) because of EQGMs. Illustrations incorporate viscoelastic (strong and liquid) dampers, liquid thick dampers, metallic-yielding and rubbing gadgets and lead expulsion frameworks. Every one of these gadgets are introduced inside the auxiliary encircling framework. A favored term for "include damping" is "include vitality scattering gadgets."
- Active control frameworks give seismic insurance by forcing powers on a structure that balance the tremor prompted powers. Innately more entangled than seismic separation or vitality dispersal frameworks (EDS), they incorporate PC controlled actuators that enact propping or tuned-mass dampers situated inside the structure.
- Hybrid control frameworks are generally a mix of dynamic and latent frameworks. Lessened power requests and enhanced unwavering quality are the fundamental framework highlights. Latent frameworks incorporate a mix of seismic isolators or EDS or both.
- Although the above definitions are utilized generally in the United States, Bertero (1997a, 1997b, 2002) in the discourse of the utilization of seismic segregation frameworks and latent vitality dissemination gadgets has proposed a few changes, which were guided by the utilization of vitality ideas.

Recent Advances in the Design and Construction of Innovative Structural Systems

Late advances in the outline and development of building basic frameworks utilizing the inventive methodology and innovation of seismic disconnection are talked about in Chapter 11. The advances and uses of the methodology and innovation of detached vitality scattering gadgets in the United States are examined in Chapter 12 and in late productions by Hanson and Soong (2001), Soong and Dargush (1997) and Constantinou et al. (1998). Section 12 additionally talks about a few applications in Japan. As said beforehand, the present section is predominantly dedicated to talking about the advancement and utilization of creative systems and procedures of EDS in Japan. These frameworks depend on the idea of DTS or DCS, which is hypothetically in light of the execution based outline system proposed by Connor and Wada (1992) and its applications utilizing different vitality scattering gadgets with specific accentuation on the utilization of "unbonded supports" or "growing limited propped outlines (BRBFs)."

USE OF INNOVATIVE STRATEGIES AND TECHNIQUES IN JAPAN

Overview of Seismic Design Trends in Japan

Japanese seismic plan guidelines characterize two levels (level 1 and 2) of EQGMs and permissible harm for each of these levels. For level 1, little and direct EQGMs, just minor harm, for example, splits in dividers and pillars are permitted, while human life and the building capacities and structure are ensured. For level 2, an extreme uncommon EQGM, a building structure might be harmed given that human life is secured. The flow seismic plan and research in Japan depend on these prerequisites. In any case, since the structures have as of late expanded in their sizes, and furthermore in light of the need to oblige costly PC and correspondence gear, and in addition the usefulness of such hardware, the above ordinary plan necessities should be adjusted. The lessons gained from the Northridge and Hyogoken-Nanbu quakes accentuate the need to perceive that the harm to the auxiliary and nonstructural frameworks planned by the past rationality may bring about incredible human life and monetary misfortunes. Clearly too substantial plastic distortion in basic and nonstructural frameworks of a building ought not be took into account extreme EQGMs. Moreover, development exercises requiring the creation of concrete and steel raise new worries about natural issues, for example, destroying precipitation timberlands and expanding CO_2 The seriousness of these issues could be diminished by stretching the building's administration life. In this way, it turned out to be evident that another outline approach was required with the end goal that structures, particularly huge vital structures, stay practical after direct EQGMs as well as even after a serious occasion. A promising methodology seems, by all accounts, to be in the utilization of the idea of "harm tolerant" basic framework, which is portrayed next.

The idea of a "harm tolerant" structure was proposed in Japan before the Northridge and the Kobe seismic tremors (Wada et al., 1992). Harm tolerant implies that the satisfactory harm because of a seismic tremor happens in particular basic segments, (for example, props, shear dividers or supplemental vitality disseminating gadgets) that are added to what is called "essential basic framework." These harmed segments are known as the "relinquishing individuals" and capacity to some degree like a circuit to shield the essential structure from extreme harm. After the 1994 Northridge and 1995 Hyogoken-Nanbu seismic tremors, "harm tolerant" or DCS have gotten expanding consideration by scientists and basic specialists particularly in the United States and Japan. Accordingly, it is not astonishing that the examination of the seismic conduct of DCS and its applications in the outline and development of structures situated in areas of high seismicity have broadly expanded around the world (see Chapter 16) (Bertero, 1996, 1997a, 1997b; Housner et al., 1997; Huang, 1995; Huang et al., 1994; Soong and Constantinou, 1994; Soong and Dargush, 1997; Wada et al., 1992, 1997, 1998; Wada and Huang, 1995).

On the front of the Engineering News-Record (Reina and Normile, 1997), "conciliatory" was utilized together with an applied photo of a DCS and a short article. The article clarifies that the vitality ingestion that occurred from the hub yielding of the damping props (they call conciliatory supports) that were utilized as a part of the steel basic framework will turn into the "relinquished" props through which high, rise building structure will survive even a serious EQGM. As is talked about later in this section, in extraordinary minute opposing edges (SMRFs) the welded spine of the shaft closes progress toward becoming yielded. Figure 7.4 demonstrates a draw of a run of the mill pillar section association in the United States before the Northridge tremor.

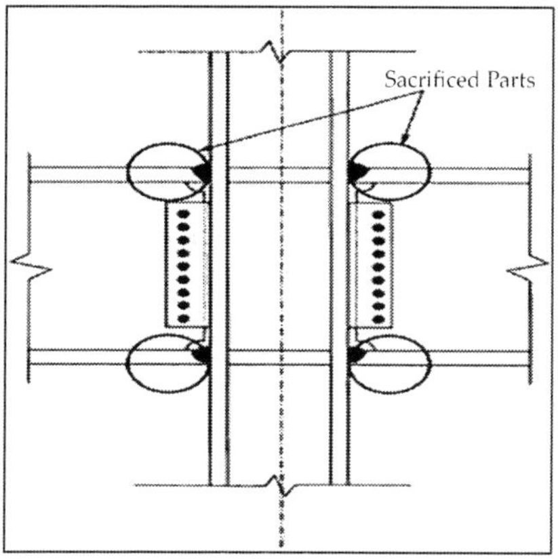

Figure7.4: Typical beam-column connection in the United States before Northridge earthquake

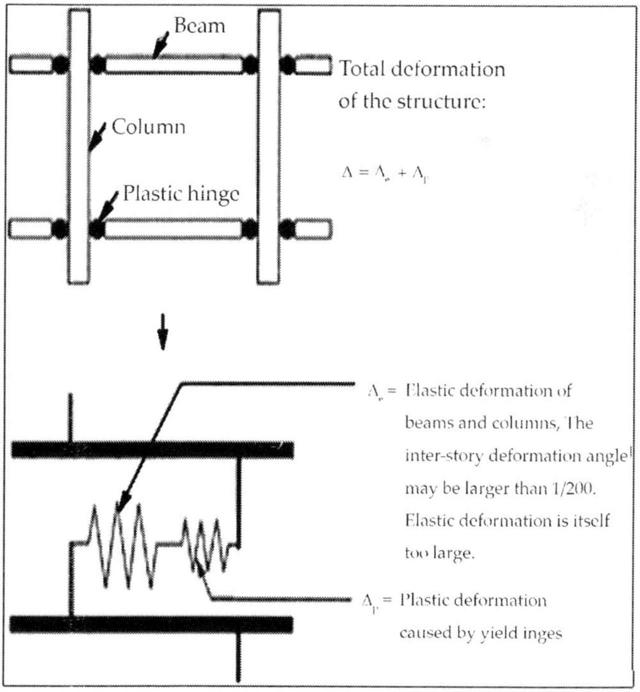

Figure 7.5: Strong column-weak beam model.

Little vitality dissemination can be normal from the plastic misshapening of these sorts of shaft finishes amid an extreme EQGM

(Bertero, 1980) (as was obviously exhibited amid the Northridge [Engelhardt, 1995, Carper, 1998, Krawinkler, 1994] and the Hyogoken-Nanbu seismic tremors [Steel Committee of Kinki Brach, 1995]) in light of the fact that the plastic distortion of the bar closures is identical to the strategy for mounting elasto-plastic vitality scattering gadgets in arrangement with a versatile edge (Figure 7.5), prompting huge disfigurement of the entire edge after it ends up plainly plastic instrument, which requests noteworthy vitality dispersal that the welded associations of the ribs can't supply.

Comparison of U.S. and Japanese Seismic Design Criteria

SMRF

Both the United States and Japan have similar procedures for the design of SMRF structural systems. These procedures are based on consideration of a minimum required strength and maximum acceptable deformation.

TABLE 7.1: Comparison of US-Japanese Seismic Design for MRSF Structure

	Japan	USA
Building height≤60		Base shear
m		
Base shear		for any earthquake
$V \geq 0.2R_tZW$	for moderate earthquake	
$V \geq 0.2R_tZD_sW$	for severe earthquake D_s=0.25	C_v: seismic coefficient varies with soil type
	for SMRF	and location,
Design PGA:		I: importance factor of the building,
A_g=0.08g~0.1g	for moderate earthquake	T: fundamental period of the building,
A_g=0.4g~0.5g	for severe earthquake	R: Response modification factor, R=8.5 for

		SMRF.			
Deformation		Maximum inelastic response displacement:			
Criteria:					
	for moderate earthquake	Δ_M=0.7 RΔ_S			
Δ_M is not limited for severe earthquake				Story drift limit:	
Building height>60					
m					
Time history analysis have to be done.					
Design PGV:				Design level defor-mation Δ_s for MRSF	
V_g=25cm/s (A_g=0.25g) for moderate earthquake					
V_g=50cm/s (A_g=0.50g) for severe earthquake					

The required strength is defined by the base shear and the required maximum deformation is defined by the maximum acceptable inter story drift. However, there are differences in the quantifications of these requirements. Table 7.1 lists some typical specifications.

In Japanese seismic design specification, the inter-story drift Δs (for T greater than or equal to 0.7 sec) is limited to be less than hs/200. In the U.S. seismic specification, however, the inter-story drift Δs is limited less than hs/298 (for T greater than 0.7 sec), which is why the depth of structural members of the United States is usually designed to be larger than those of Japan. Some detailed discussions of the above specifications are given in the following sections.

Seismic Design Criteria for SMRF in Japan

Japanese seismic outline techniques for building structures are diverse relying upon whether the building stature is past or under 60 m (197 ft).

For the structures whose tallness is under 60 m (197 ft), flexible plan is required utilizing base shear coefficient of 0.2, which is related with little to direct EQGM. The Inter-story float edge must be under 0.5%. The most extreme increasing speed of the ground movement is thought to be 80 to 100 cm/sec2. For extreme EQGMs whose most extreme ground movement increasing speed is 400 to 500 cm/sec2, five times bigger base shear coefficient (i.e., 1.0) ought to be utilized for a definitive state outline. This requested shear constrain could be decreased relying upon the distortion limit of structure. For the most malleable steel SMRF, the predetermined Ds element is 0.25. For strengthened solid divider structures, which have less pliable limit, the Ds variable is 0.55. In any case, there is no predetermined misshapening limit for a definitive state configuration in the event of an extreme EQGM.

Since the underlying versatile firmness of minute opposing steel casing is little, the cross-sectional measurement of most the SMRF structures is controlled by the predefined Δs=0.5%hs under the base shear coefficient of 0.2. Since extreme EQGMs are considered to have top ground speeding up Ag equivalent to five times those predetermined for the little EQGM, the greatest interstory float edge under the serious EQGM is thought to be 1/40. Besides, as the plastic distortion creates, bringing about an abatement in solidness, the interstory float edge could achieve 1/30. This was affirmed by perceptions of the harmed steel SMRF after the Hyogoken-Nanbu tremor in 1995. Time history reaction examination is required for the seismic outline of tall structures more than 60 m (197 ft) high. For the little and direct EQGMs, the most extreme speed of ground movement is considered as 25 cm/sec or the related greatest increasing speed is roughly 250 cm/sec2. The interstory float point must be controlled to values under 0.5%. For an extreme EQGM, the greatest speed of ground movement is considered as 50 cm/sec or related most extreme quickening is 500 cm/sec2; the interstory float point is required to be under 1.0%. The fundamental contrasts in seismic outlines between more than 60 m (197 ft) elevated structures and under 60 m (197 ft) low-ascent structures are (1) the extent of the most extreme speeding up of little and direct EQGMs utilized for tall structures is around 2.5 times bigger than low-ascent structures and (2) while there is twisting breaking point to the tall working for serious EQGMs there is no restriction for the low-ascent structures. From these prerequisites it is obvious that in Japan there are more stringent necessities for tall structures than those for low-ascent structures(Huang et al., 2004).

Seismic Design Criteria for SMRF in the United States

An outline strategy to change the required basic quality as indicated by the plastic misshapening limits (flexibility limit) of the structure was first created in the United States. R figure esteem has been indicated in the U.S. codes since 1988 (ICBO, 1988); 10 years after it had been presented by the ATC 306 report (1978). Calculate R is equal speculation to the reverse of the Japanese Ds esteem. Flexible reaction level is figured in view of the neighborhood and basic period. The shear constrain utilized for flexible plan is gotten from the versatile reaction compel, which is partitioned by the variable R. For the malleable minute opposing steel outlines, UBC 97 (ICBO, 1997) suggests R=8.5; for less pliable fortified solid bearing divider auxiliary framework, R=4.5 ought to be utilized.

The U.S. seismic codes constrain the most extreme satisfactory inelastic reaction removal ΔM, which is not determined in Japan. As per the flexible outline system in the U.S. code, the plan yielding quality is gotten by lessening the required flexible powers by an element of R, and the most extreme versatile request uprooting Δs, ought to be duplicated by a variable 0.7.R to acquire the greatest inelastic interstory float ΔM, i.e., ΔM=0.7.RΔS. As the U.S. code determines the most extreme adequate ΔM=(2.5%)hs for structures having T<0.7 sec and ΔM≤(2%)hs for T≥ 0.7 sec and to get the greatest worthy or farthest point in the estimation of the flexible outline level reaction dislodging Δs (i.e., the story float under the plan EQGM) the above ΔM esteem ought to be increased by an element of 1/(0.7.R).

From the above exchanges it turns out to be certain that while in Japan a similar distortion restrict 1/200, related with base shear component of 0.2, is required for a wide range of structures, in the United States, as far as possible relies on upon the inelastic twisting limit of the structure. Along these lines, for malleable structures in the United States, the seismic outline shear compel utilized as a part of the predetermined direct flexible plan system can be lessened.

United States-Japan Comparison on the Design of Steel SMRFs

As examined beforehand, in Japan, for the working with tallness under 60 m (197 ft), as far as possible prerequisite is determined for little and direct EQGMs. The plan base shear for Japanese structures is bigger than that predetermined in the United States. In this way, in Japan it is sensible to plan and fabricate every one of the edges in a structure as SMRF, i.e., all with inflexible associations.

In the United States, the code indicates a distortion constrain for the outline against serious EQGMs and requires that the direct flexible plan method be founded on the lessened estimations of the base shear drives that will be required by the straight versatile reaction for the plan EQGM. The lessening component is L/R for shear strengths. The disfigurement is processed by Table 7.1.

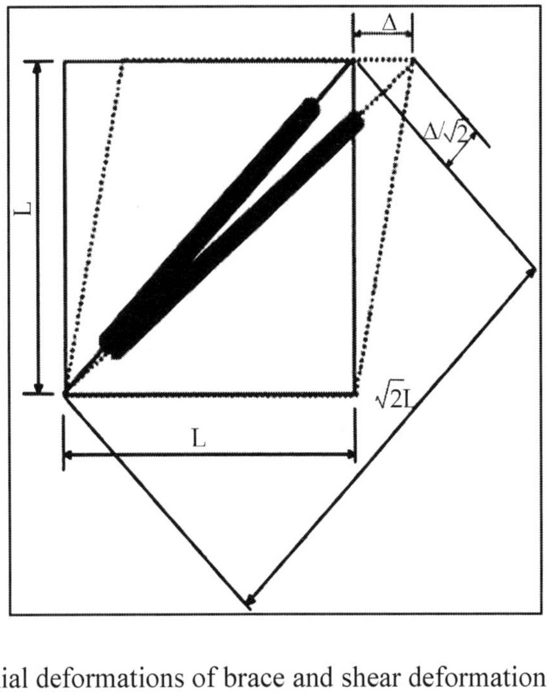

Figure 7.6: Axial deformations of brace and shear deformation of frame.

As the required outline base shear compel is littler than that in Japan, it is extremely basic to plan only couple of edges to oppose the horizontal strengths actuated by the EQGMs and every single other casing are intended to oppose the gravity loads. Since the base shear compel is little, the other edge individuals don't have to oppose the level powers, and in this manner, the shafts are associated with the sections just through web. In any case, in light of the little worthy sidelong misshapenings, the creator needs to choose bars and segments with extensive and solid cross segments for SMRFs. On this premise of the present Japanese learning, the SMRFs in the United States can be called uninvolved controlled seismic edges in which the welded associated spines of the pillars at their closures are considered as the vitality disseminating gadgets or "give up individuals."

Such sort of steel SMRFs endured serious harms amid the 1994 Northridge tremor. Subsequently, after 1994, in the United States the utilization of base disengagement and vitality dissemination gadgets (especially unbonded props) wound up noticeably alluring contrasted with proceeding with utilize few SMRFs with vast size pillars and sections. It is trusted that the utilization of these gadgets allows an extremely balanced and both actually and monetarily productive seismic plan, particularly in the United States, where the strict necessity restricting the parallel twisting can be met and satisfactory sidelong resistance limit can be gotten. Since in the above imaginative techniques it is not important to utilize overwhelming substantial size pillars and segments requested by few SMRFs, it is conceivable to lessen the aggregate sum of steel required for the whole casing structure.

Deformation of Buildings

For a working to be useful, it must have basic segments (pillars, segments and dividers), as well as segments, windows, and entryways, utility lines and gear. The last segments are called nonstructural individuals. These nonstructural parts are effortlessly seriously harmed when the interstory disfigurement of the building is substantial (Oliveto, 2002).

Give us a chance to consider the sidelong distortion of an edge comprising of a pillar and segment with a support set at 45°, as appeared in Figure 7.6. Contrasted with the sidelong shear misshapening ±Δ happening in the casing, the development and construction of the support moves toward becoming in light of the fact that the prop length is times the segment length L, the hub strain in the prop moves toward becoming 1/2 of the shear distortion point of the casing. Considering the way that the joints at the prop finishes are of high firmness and that they will remain

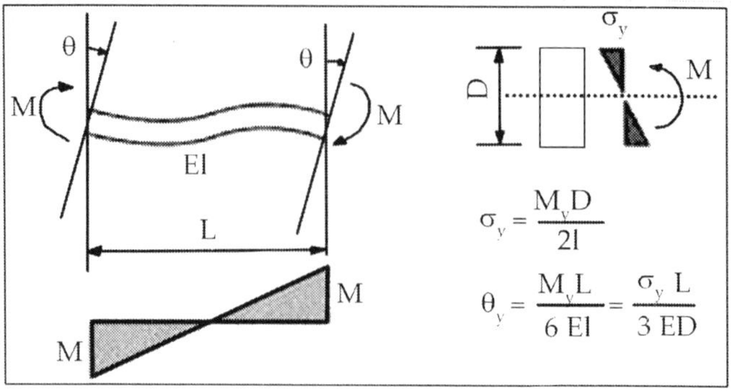

Figure 7.7: Deformation angle at beam ends of a rigid frame

elastic and assuming that the yield strain of the steel brace is 0.1%, it is found that when the brace yields, the story deformation angle is as small as 1/500.

For a steel plate shear wall, according to Von Mises's plastic theory, the yield shear stress of steel is $1/\sqrt{3}$ times the yield normal stress. The shear elastic modulus is 1/2.6 times Young's modulus. The shear yield strain is then around 1.5 (=2.6/1.732) times the hub yield strain. It implies that the story twisting edge winds up noticeably around 1/667 when the steel plate shear divider starts to yield and it ends up noticeably plastic at nearly an indistinguishable level from the little story disfigurement plot for the support. It is conceivable to make the story twisting edge toward the start of plastic disfigurement littler for the prop and steel plate divider by utilizing ultra-low yield steel whose yield quality is in the vicinity of 100 and 200 MPa. Give us a chance to consider the yield disfigurement point of an inflexible joint edge involving segments and shafts, which get a lopsided bowing minute (Figure 7.7). The aggregate parallel distortion of a casing structure comes about because of the impacts of the accompanying five segments: twisting misshapening and shear disfigurement of segments, bowing twisting and shear misshapening of pillars, and shear twisting of the board zone in shaft section joint. In a general building outline structure, the segment area is typically bigger than the bar segment, and the segment tallness is normally shorter than the pillar traverse. Along these lines, the section twisting is substantially littler than the pillar distortion. The bar bowing disfigurement contributes roughly half of the aggregate twisting of the entire casing structure. For the steel structure outline, discourse is centered around the rotational mutilation happening at the shaft closes. Let the span and the depth of the beam be L and D, respectively. Then, the deformation angle θy at beamends, when the stress of the flange reaches the yield point σy becomes (σy/3E)(L/D) (see Figure 7.7).

$$\theta_\gamma = \frac{M_\gamma L}{6EI} \qquad (7.5)$$

$$\sigma_\gamma = \frac{M_\gamma D}{2I} \qquad (7.6)$$

$$\theta_\gamma = \frac{\sigma_\gamma L}{3ED} \qquad (7.7)$$

The traverse L of the edge is foreordained and Young's modulus E is a consistent of preselected material. Along these lines, the distortion at the yield purpose of the casing can be expanded by utilizing steel having a high return point σy and individuals having a littler profundity D than customary ones. At the end of the day, the versatile twisting limit of an edge can be expanded utilizing a thin adaptable casing fabricated by high strength steels. Subsequently, the yield disfigurement existing apart from everything else safe casing can be effectively dictated by choosing the materials and the profundity of the auxiliary individuals. In actuality, the yield misshapening of the controlling individuals, for example, supports or shear dividers is resolved from the general design of the edge and the yielding quality of the chose material. Hence, the measure of plastic distortion can›t be expanded by just altering the plate thickness and itemized arrangement.

At the point when the plastic distortion of the controlling or vitality dispersing individuals are utilized as a part of blend with the occasion opposing edge, ignoring issues, for example, support clasping, shear disappointment of shear divider, and so forth., the accompanying challenges still occurred. Plastic disfigurements happen first in each of the controlling individuals (or areas of these individuals) as opposed to at the time opposing casings, so firmness and quality crumbling of the controlling individuals (props or dividers) happens, while the quality of the edges keeps on expanding. Accordingly, it is hard to have a dependable estimation of the aggregate reaction and quality of the basic frameworks.

The innovation of DCS is hypothetically established on movement based outline and is reliable with the idea of execution, based plan (Connor, 2003; Connor et al., 1997; Connor and Klink, 1996; Connor and Wada, 1992). This innovation is not constrained in the plan of building structures, even in the outline of modern offices at present, there is a pattern to discover better plan factors to meet different outline prerequisites. Much of the time, the plan necessities must be changed amid the long administration time of the structure. The most critical methodology of planning a structure is to rearrange however much as could be expected the outline prerequisites and plan factors. The outline system of DCS depends on a straightforward idea. The capacity of the building structure has two clear plan necessities. One is to oppose vertical gravity stack, another is to oppose the impacts of EQGMs. In the outline procedure, the entire building structure can be partitioned into two individual plan factors (auxiliary frameworks). One is the essential structure, for example, shafts and sections, which are intended to oppose just the vertical administration stack; the other is the controlling or vitality disseminating framework, which is intended to oppose the horizontal strengths coming about because of the impacts of EQGMs(Huang et al., 2004).

Applications of EDS

Building structures must be intended to have the capacity to withstand greatly vast misshapenings that can happen arbitrarily over the administration life of the building. The seismic tremor debacles specified beforehand highlight the need to outline and manufacture solid and versatile structures have the capacity to withstand the impacts of extreme EQGMs with controlled harm to levels that will enable the capacity of the working to be reestablished at the earliest opportunity after such serious EQGMs. Along these lines, in quake safe plan of building structures, despite the fact that the need ought to be given to assurance of human lives against extreme uncommon EQGMs, it is additionally important to control the level of auxiliary and nonstructural harm to maintain a strategic distance from long intrusion of the capacity of the building, since this can bring about huge financial misfortune. This issue has been viewed as genuinely by many individuals in the United States and Japan as well as in different nations.

As talked about beforehand, the customary seismic outline of SMRF structures depends on the presumption that under serious EQGMs the districts at the finishes of the shafts can grow adequately substantial plastic misshapening to disseminate a critical segment of the tremor input vitality. Accordingly, this traditional approach for SMRFs, which has been designated as the "solid segment frail shaft" plan, has so far been acknowledged and connected worldwide by auxiliary architects. Figure 7.5 demonstrates the idea driving solid section, powerless shaft SMRF structures. Such structures can be dealt with as an arrangement of two springs that are associated in arrangement. The aggregate disfigurement of such structure after the pillar end locales yield is the summation of the flexible distortion Δe and the plastic misshapening of the plastic pivots Δp. Clearly, the plastic pivots at the shaft closes increment the aggregate distortion of the whole structure and lessen the sidelong solidness of the entire building structure. In the United States, the greatest aggregate misshapening is constrained by the prerequisite that the aggregate shear or interstory disfigurement edge ought not surpass the estimations of 0.020 or 0.025, contingent upon the auxiliary time frame.

After the Northridge and Hyogoken-Nanbu seismic tremors, it was watched that many steel SMRF structures were harmed, in spite of the fact that they had been viewed as the perfect auxiliary framework to oppose the impacts of EQGMs. The fundamental harm has been the weak crack at the welded area between the beamends and segments (Carper, 1998; Engelhardt and Sabol, 1995; Investigation Committee of Building's Damage, 1995; Krawinkler, 1994; Nakagomi, 1997; Nakagomi and Oka, 1996). In this way, the act of depending on the essential SMRF structure to retain and disperse a noteworthy part of the info vitality of EQGM in the end brings about serious harm to these essential structures. So also, from an efficient perspective, on account of strengthened cement SMRF structures, the regular practice has been simply to depend on the plastic disfigurement at shaft finishes to scatter a lot of the vitality contribution from the seismic tremor excitation.

Concept of DCS

As pointed out beforehand, the essential idea of DCS (Wada et al., 1992, 1997) can be portrayed as takes after. The incorporated whole building auxiliary framework is the mix of two distinct structures, as appeared in

Figure 7.8: One is the essential structure made out of pillars and segments, which intends to oppose the vertical administration stack.

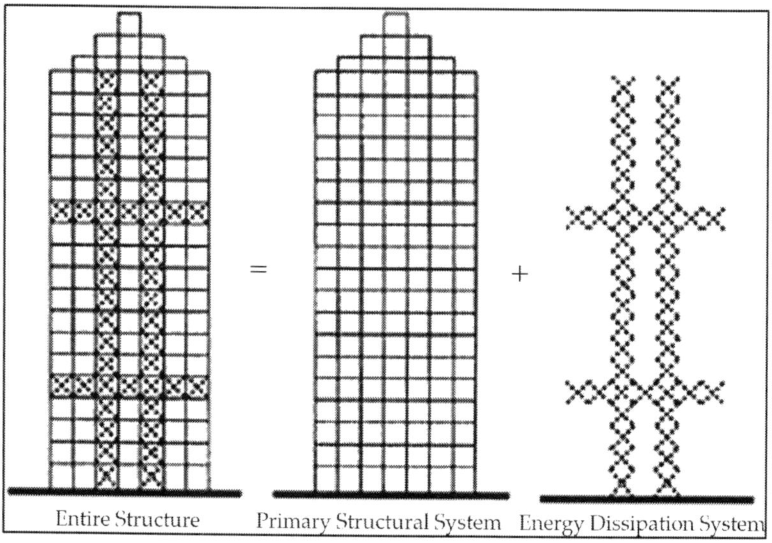

Figure 7.8: illustrates the structural modeling of a DCS.

The essential structure is intended to carry on flexibly and to hold its building administration works notwithstanding amid an extreme EQGM. The second is the vitality scattering or harm controlling framework that expects to control the impacts of the horizontal strengths and disfigurements coming about because of the EQGM. Along these lines, the harm actuated by the EQGM is controlled by this vitality dispersal framework, which is effortlessly checked and repaired or supplanted after an extreme EQGM.

The harm is controlled inside the support sort vitality scattering framework. The essential auxiliary casing and the harm controlling framework can be viewed as an arrangement of two springs associated in parallel. The aggregate distortion of the whole structure is equivalent to the edge disfigurement Δf and furthermore equivalent to the misshapening Δd of the vitality disseminating framework. The benefit of this auxiliary framework is to have a more solid vitality scattering framework with an expanded vitality dispersal limit, firmness and quality of the essential structure without expanding the aggregate disfigurement of the whole structure.

Figure 7.10 analyzes the regular solid section feeble pillar structures and the DCS structures with a vitality dispersal framework (Onishi et al., 1997). For little and direct EQGMs, the customary structures are intended to stay flexible. This implies the common auxiliary damping is the main instrument that disseminates some portion of the info vitality of these little and direct EQGMs. In this manner, if the characteristic auxiliary damping is overestimated amid the plan arrange, the essential structure will be subjected to expansive distortion and drive or yield even under little and direct EQGMs

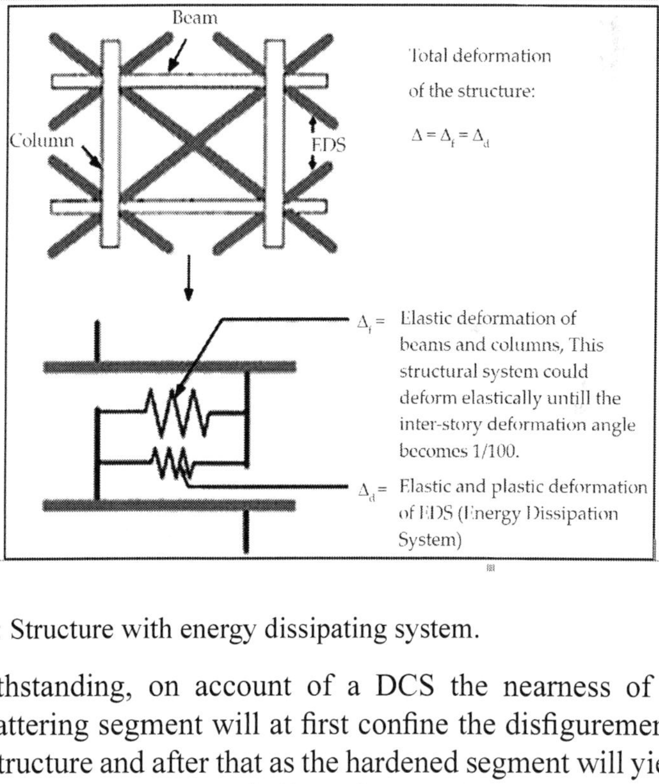

Figure 7.9: Structure with energy dissipating system.

Notwithstanding, on account of a DCS the nearness of the firm vitality scattering segment will at first confine the disfigurements of the essential structure and after that as the hardened segment will yield under little parallel distortions, it will disseminate a piece of the EQGM input vitality and it will keep controlling the misshapening of the essential basic framework.

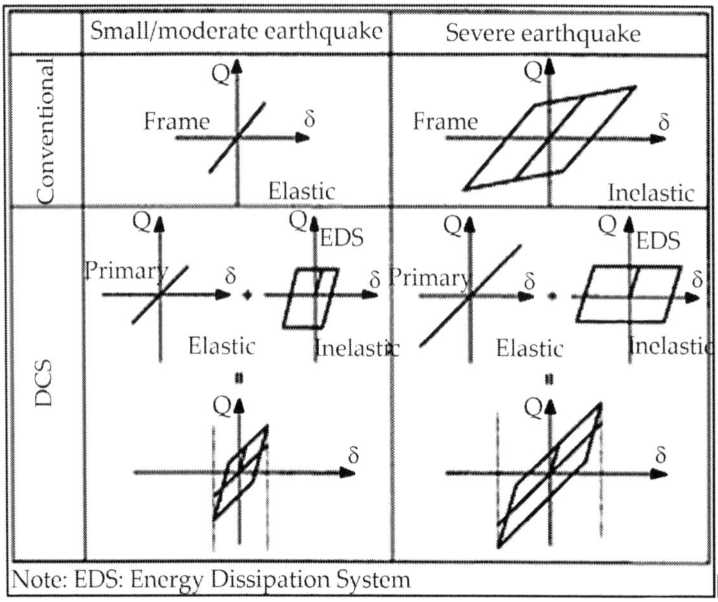

Figure 7.10: Comparison between conventional and DCS structural models.

Under extreme EQGMs, the regular SMRF structure depends all alone yielding ability to scatter the information vitality, and along these lines, will maintain vast plastic distortions, which implies a lot of harm. The bigger the requested plastic distortions (physical pliability) the bigger will be the harm of the whole SMRF framework. Then again, if a DCS framework is utilized, it will be conceivable to control the reaction of its essential basic framework with the end goal that it stays versatile, in light of the fact that the vitality scattering segment or framework will build the solidness, quality and especially the vitality dispersal limits of the whole basic framework.

The benefit of the DCS is not just in shielding the essential structure from harm amid an extreme EQGM, additionally in diminishing the development cost when contrasted and that of regular SMRF. As indicated by the report of Nikkei Architecture, "steel structures planned as DCS the aggregate weight of the steel can be decreased by 20%" (Nikkei, 1997). Besides, as it has been beforehand brought up, maybe a more imperative favorable position of DCS is that after an extreme EQGM if its vitality dispersal parts have been harmed, they can be immediately expelled and supplanted.

Mechanism of Unbonded Braces

The unbonded prop is an exceptional kind of vitality dispersal gadget that was created by Nippon Steel in Japan, 1987. The unbonded prop is a perfect steel part that is utilized as the steel center plate bar compelled against clasping by a solid filled steel tube.

A kind of extraordinary bond-breaking material, which is known as the unbond material, is utilized between the solid and center steel plate to decrease the rubbing activity between the steel center plate and the solid. Hence, the outside steel tube and cement are not subjected to the impacts of the pivotal part compel that must be opposed by the unbonded prop. On account of this kind of manufacture, the unbonded support can have a to a great degree stable hysteretic conduct under both tractable and compressive powers. The firmness and quality of a building can be changed in accordance with any coveted an incentive by the ideal determination of the cross-segment territory and quality of center steel plate.

It has been affirmed from many test outcomes (Iwata, 1995; Iwata et al, 1995; Wada, 1988) that the unbonded support shows stable hysteresis if the clasping quality (Pcr) of simply the steel tube is higher than the result of a security consider fs and the yield quality (Py) of the center steel plate.

$$P_{\alpha} = \frac{\pi^2 EI}{L_k^2} > f_s P_{\gamma} \tag{7.8}$$

where P_y is the yield strength of the core steel plate and f_s is the safety factor (1.5 is recommended). I is the geometrical moment of inertia of the steel tube, and L_k is the unbraced length of the unbonded brace. It should be noted that the actual buckling restrained capacity of an unbonded brace depends only on the bending stiffness of the exterior steel tube and the concrete, i.e., it has nothing to do with the material strength and cross section of the core steel plate.

Figure 7.11 demonstrates the test setup for leading stacking cycles on a steel outline just and another steel outline with unbonded support. The casings were made of high-quality steel. The outcomes appeared in Figure 7.12 allow examination of the hysteretic conduct of such sort of steel edges, i.e., with and without unbonded props. Figure 7.12a demonstrates the compel uprooting hysteretic circles for the steel outline alone (MRF1). The steel outline itself yields and, on the grounds that it scatters little vitality as a result of plastic misshapenings that were endured by the casing itself, in this manner makes harm the casing. Figure 7.12b demonstrates that the steel outline with unbonded prop (MRF2) through its expansive hysteretic circles can scatter a lot of vitality. The steel outline MRF1 was composed with an indistinguishable quality limit from the MRF2 in addition to support. As all the plastic strain vitality is scattered by the unbonded support, the reaction of the steel outline remains nearly in its versatile range. In this way, the steel casing is not harmed(Huang et al., 2004).

The unbonded support has numerous great components when it is legitimately utilized as a kind of vitality dispersal gadget. Some run of the mill focal points of utilization of the unbonded prop are outlined as takes after:

- o Simple displaying of its conduct for its auxiliary investigation
- o Easy treatment of its erection and amassing it to the development of the basic framework
- o Stable hysteretic conduct without clasping
- o Design adaptability in the determination of solidness and quality of the whole auxiliary arrangement of a building

There are, nonetheless, regions that should be additionally considered with respect to the execution and harm of the unbonded support framework, including:

- Lack of recentering limit and measure of perpetual disfigurement
- Checking the harmed props and criteria for their substitution
- Effects of three segments of ground movement
- Study of unwavering quality of association of the props with the casing

Applications of EDS in Tall Buildings in Japan

Since the Hyogoken-Nanbu seismic tremor, many building ventures that were composed in light of the idea of DCS have been modified by the Japanese Building Center. Table 7.2 gives a rundown of such normal tasks that were planned and built in the vicinity of 1995 and 1998. Since 1999, roughly 70% of the elevated structures with statures more than 60 m were planned in light of this idea of DCS.

Central Government Building

The Central Government Building (Figure 7.13) situated in Chiyoda-ku, Tokyo, is an average DCS joining hysteretic vitality dissemination of steel shear dividers and thick liquid damper. This building was composed by the Architecture Department of the Ministry of Construction and Kume Sekkei Co., Ltd. The aggregate stature is 144.5 m (474.1 ft), including a 55 m receiving wire tower on the rooftop. The superstructure over the ground level is a minute opposing steel outline introduced with different EDS, while the underground structure is a steel fortified solid casing with strengthened solid shear dividers.

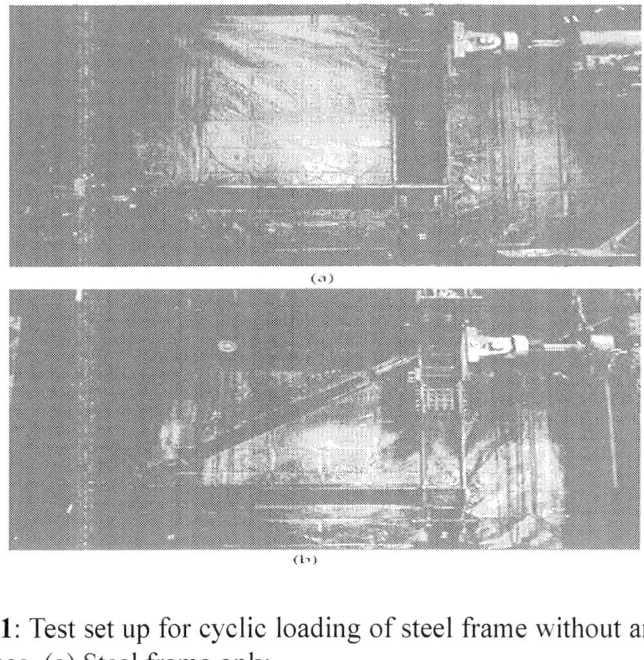

Figure 7.11: Test set up for cyclic loading of steel frame without and with un-bonded brace, (a) Steel frame only

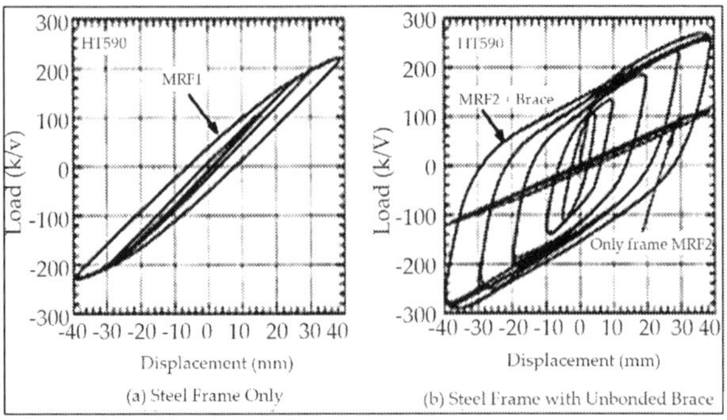

Figure 7.12: Hysteretic behavior of steel frames without and with unbonded brace, (a) Steel frame only;

Figure 7.13: Central Government

TABLE 7.2 Tall Steel Buildings Designed Based on the Same Concept of DCS between 1995 and 1998

Month, Year	Project Name	Location	Usage	Height (m)	Structure Type	Dampers	Ductility Ratio of Primary Frame
June 1995	International Congress	Osaka	Congress	104	S_F	HD_B	0.95
July 1995	Todai Hospital	Tokyo	Hospital	82	S_F	VD_S	0.93
July 1995	Tohokudai Hospital	Sendai	Hospital	80	S_F	VD_S	0.97
Aug. 1995	Central Government	Tokyo	Office	100	S_F	HD_S+VD_S	0.78
Oct. 1995	Harumi 1 Chome	Tokyo	Office, shop	175	S_F	HD_B	0.88
Feb. 1996	Toranomon 2 Chome	Tokyo	Office, shop	94	S_F	VD_S	0.94
Mar. 1996	Sankyo	Tokyo	Office	61	S_F	HD_B	0.88
Apr. 1996	Shiba 3 Chome	Tokyo	Office	152	S_F	HD_B	0.97
June 1996	Art Hotel	Sapporo	Hotel	96	S_F	HD_BD	0.85

Aug.	Kanto Post	Saitama		Office	130	S_F		VD_S	0.87	
1996	Office									
Oct.	Nakano	Tokyo		Office, shop	96	S_F		VD_S	0.68	
1996	Urban									
July	DoCoMo	Tokyo		Communica-tion,	240	S_F		VD_S	0.79	
1997	Tokyo			etc.						
Oct.	Minato	Yokohama		Hotel, shop,	99	S_F		HD_BD	0.98	
1997	Future			office						
Nov.	Nishiguchi	Yamagata		Office, hotel,	110	S_F		HD_B	1.00	
1997	Shintoshin			etc.						
Feb.	DoCoMo	Nagano		Communica-tion	75	S_F		VD_S	0.89	
1998	Nagano									
Apr.	East Osaka	East		Office	120	S_F		HD_S	1.00	
1998	City	Osaka								
May	Kouraku	Tokyo		Office, Shop	82	S_F		HD_B	1.00	
1998	Mori									
July	Harumi 1	Tokyo		Office, shop,	88	RC_F		HD_B	1.00	
1998	Chome			etc.						
Nov.	Adago 2	To-kyo		Office, shop	187	S_F		VD_B	0.71	
1998	Chome									

Nov.	Gunyama		Fu-ku-shim	Shop, school,	128	S_F	HD_B+VD_S	0.98	
1998	Station		a	etc.					

Figure 7.14: Art Hotel in Sapporo

The segments and light emissions essential structure utilized SN490B steel (most extreme quality is 490 MPa or 71 ksi) . The essential structure is intended to carry on flexibly even under a serious EQGM whose most extreme ground speed is 50 cm/sec (relating to Ag of 500 cm/sec2). The greater part of the tremor vitality is assimilated and scattered by the vitality dispersing framework. The hysteretic dampers (HDs) are steel dividers made of additional low yield point steel (yield point is 100 MPa or 14.5 ksi). The yield shear compel level of HDs at the principal floor area is thought to be 5% of the aggregate building weight. The appropriation of yield shear drive all through the stature of the building is thought to be relative to the circulation of yield shear compel of the essential structure. Then again, the gooey dampers (VD) comprise of two mobile steel plates swung from the upper bar and three settled steel plates remained on the base bar. The space between the mobile steel plates and the settled steel plates is loaded with thick fluid like silicone oil.

Art Hotel in Sapporo

The Art Hotel in Sapporo (Figure 7.14), planned and developed by Kumagai Corporation, is another harm controlled building. This building is 96 m (315 ft) high and has 26 stories over the ground. The essential structure is a minute opposing steel outline that is intended to bolster just the vertical administration stack. Two a large number of opening steel dispersal (SSD) whose shape, appeared in Figure 7.15, made of mellow quality steel (SN490B) with a yielding quality of 325 MPa or 47.14 ksi were introduced in the building. Amid a seismic tremor shaking the SSD are subjected to shear distortions through the top and base jolts and thus each slim bar encounters twisting disfigurement and begins yielding rapidly toward the end parts of the bar, even under little shear misshapening, in light of the fact that the cross area of each bar is little. Since the yielding parts of this sort of vitality dissemination are effortlessly focused on the little end parts of the thin bars, the openings ought to be produced painstakingly to maintain a strategic distance from inordinate neighborhood strain fixation. Since one bit of SSD is light to the point that somebody could hold it in one hand, the greatest favorable position is that the harmed SSD can be effectively supplanted after a serious tremor.

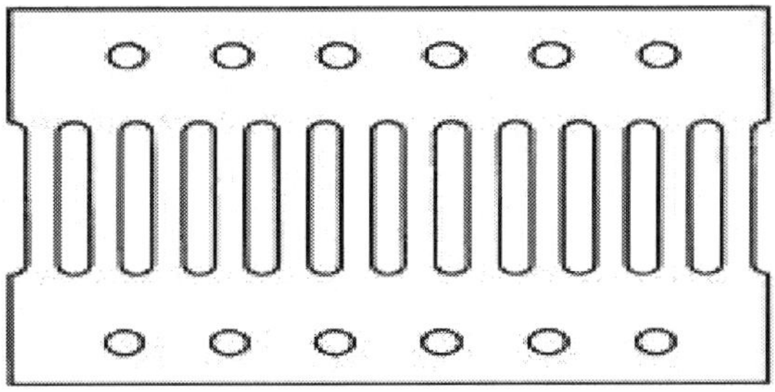

Figure 7.15: Details of the steel slit damper (SSD).

The building appeared in Figure 7.16 is situated in metropolitan Tokyo. The auxiliary outline was made by NTT Power and Building Facilities, Inc. This building has two sections. The lower 27 stories are for the most part utilized for workplaces, and on the upper 23 stories is the radio wire that is utilized for portable correspondence. There are additionally three stories underground. The aggregate tallness of the building is 240 m. The auxiliary arrangement of the upper 23 stories (the radio wire framework) comprises of a steel edge and steel support structure. The structure of the lower 27 stories is steel outline with 76 gooey damping shear dividers in both X and Y bearings. The supplemental thick damping divider framework has a similar vitality dissemination limit as that of a characteristic auxiliary damping proportion of 5% in both headings. The gooey damping divider is a high caliber, profoundly stable damping framework that has been utilized as a part of more than 10 tall building ventures since it was first utilized as a part of a seismic working in 1988. On account of the utilization of an extra gooey damping divider, the essential steel structure is intended to stay flexible even under a level-2 in number EQGM with a most extreme speed of 50 cm/sec. After basic comments about the need to utilize inventive methodologies, innovations and materials in quake building, this section was dedicated to examining the present pattern in Japan for the seismic plan of SMRF structures, that is, to construct it with respect to the idea of a DCS rather than simply the customary solid segment powerless bar SMRF framework. The essential idea for such a plan is to consider the incorporated working as a blend of two diverse auxiliary frameworks: the versatile essential structure that is outlined fundamentally to oppose flexibly the vertical administration loads, and the vitality dissemination basic framework or part to oppose the impacts of parallel EQGMs. The harm brought about by the tremor is controlled by the vitality dissemination framework. The essential structure stays versatile notwithstanding amid a to a great degree huge EQGM. In this section, the idea and logic of the DCS were surveyed. Comes about acquired from some static cyclic stacking trial of models of the primary piece of a steel SMRF and a harm controlled steel edge were displayed. The outcomes plainly demonstrate the upsides of the DCS over the regular SMRF. Chosen genuine building ventures planned on this idea were talked about, and also their favorable circumstances and detriments. In Japan right now, the idea of DCS is being connected just to elevated structures more than 60 m (197 ft) in stature. It has still not yet spread to lower-rise structures (under 60 m [197 ft]). This is for the most part a direct result of a decide that the steel structure must be planned inside the versatile locale for the base shear coefficient of 0.2 related with a direct quake. Since this manage has been acknowledged for over 50 years—since the apocalypse War II—it is not easy to change it. As said in this part, the prop and shear divider starts to yield when the interstory disfigurement edge is around 1/500. Since the essential structure can be kept in its versatile range until the interstory distortion achieves 1/200, it is sensible to permit the supports or shear dividers or both to yield. In 2004, the seismic tremor safe plan strategy for structures in light of vitality adjust, which was proposed by Professor Hiroshi (Akiyama, 1985, 2002), will be incorporated into the National Building Design Law of Japan. As indicated by this building outline law, the plastic disfigurement vitality dissemination will be took into consideration the seismic individuals notwithstanding for a little and direct quake level. In this manner, 2004 will be the year when Japanese seismic outline philosophy definitely changes. Finally, the use of the idea of harm controlled seismic plan has as of now spread to the substructure of extensions in both Japan and the United States. It is relied upon to generally utilized by the seismic-touchy nations around the world.

EARTHQUAKE RESISTANT STRUCTURES BY PLANNING AND DESIGN APPROACH

Seismic tremors have tormented man for centuries. It is a damaging power, which was sometime in the distant past pronounced to be fury of God for betrayal of people. Be that as it may, today, we comprehend what causes tremors, and can outline viable systems to alleviate the impacts of quakes. Essentially, there is the Conventional way to deal with accomplishing quake resistance, then there is the fundamental approach, and these days, there are Active Control Devices which can balance the impacts of tremors on structures.

Conventional Approach

Configuration relies on giving the building quality, solidness and inelastic distortion limit which are sufficiently awesome to withstand a given level of tremor created drive. This can be expert by choice of a proper auxiliary arrangement and watchful specifying of basic individuals, for example, bars and sections, and the associations between them.

Basic Approach

Configuration relies on basic more propelled procedures for seismic tremor resistance is not to fortify the building, but rather to decrease the quake created strengths following up on it. This can be proficient by de-coupling the structure from seismic ground movement it is conceivable to lessen the tremor instigated compels in it by three ways.

1. Increase common time of structures by Base Isolation.
2. Increase damping of framework by Energy Dissipation Devices.
3. Mitigate seismic tremor impacts totally by utilizing Active Control Devices.

ENERGY DISSIPATION DEVICES FOR EARTHQUAKE RESISTANT BUILDING DESIGN

Another approach for controlling seismic mischief in structures and improving their seismic execution is by presenting Seismic Dampers set up of fundamental parts, for instance, corner to corner props. These dampers exhibition like the water fueled shields in cars – an extraordinary piece of the sudden mongrels are put resources into the weight driven fluids and simply little is transmitted above to the body of the auto. Exactly when seismic imperativeness is transmitted through them, dampers ingest some bit of it, and in this way wet the development of the building.

Seismic Dampers

Seismic Dampers are utilized as a part of place of auxiliary components, similar to inclining props, for controlling seismic harm in structures. It somewhat ingests the seismic vitality and lessens the movement of structures.

In seismic structures overhauling, one of the horizontal constrain diminishment brought about by the quake is utilization of dampers. Amid a tremor, high vitality is connected to the structure. This vitality is connected in two sorts of active and potential (strain) to structure and it is consumed or amortized. On the off chance that structure is free of damping, its vibration will be ceaselessly, however because of the material damping, vibration is diminished. Input energy caused by earthquake to structure is presented in the following equation:

$$E = Ek + Es + En + Ed \qquad (1)$$

In this equation, E is earthquake input energy, Ek is kinetic energy, Es is reversible strain energy in the elastic range and Eh is the amount of wasted energy due to inelastic deformation and Ed is the amount of amortized energy by additional damper.

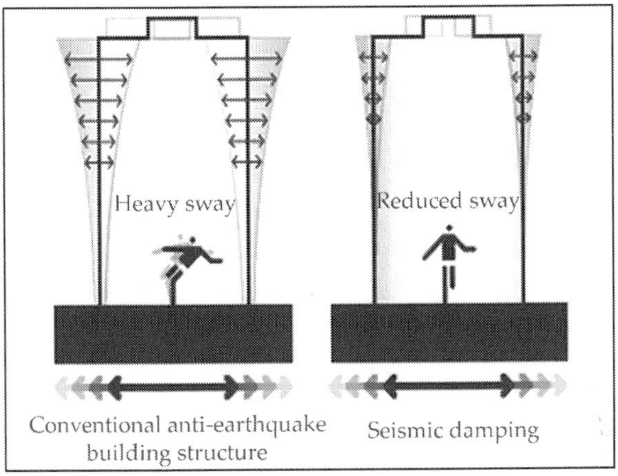

Figure 7.16: Seismic Structures Overhauling

In seismic isolation systems, use of energy dissipation systems, allocated a special place to their selves. Damping increasing is possible by using various methods such as the flow of a soft metal, two metal friction on each other and a piston motion within a slimy substance or viscoelastic behavior in materials such rubber-like substances.

DAMPING EFFECT ON STRUCTURAL RESPONSE

Damping increasing reduces structural response (acceleration and displacement) damping effect at low frequency (close to zero) have no effect on spectrum amount and at high frequency, it has low effect on response acceleration. Figures 1 and 2 show the most effect of damping increasing in the frequency of 0.3 to 2.5 seconds.

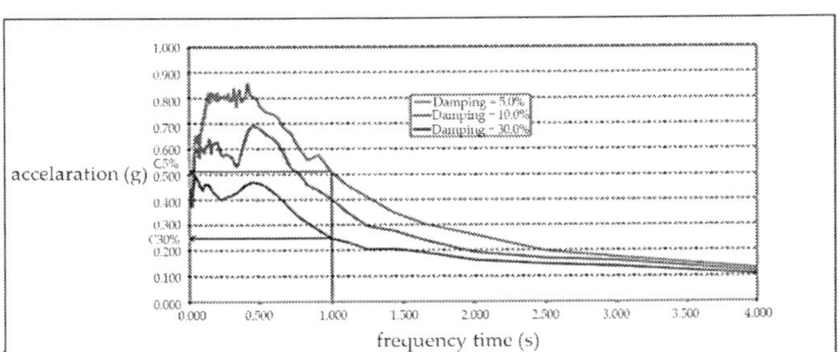

Figure 7.17: Damping Increasing Reduces Structural Response

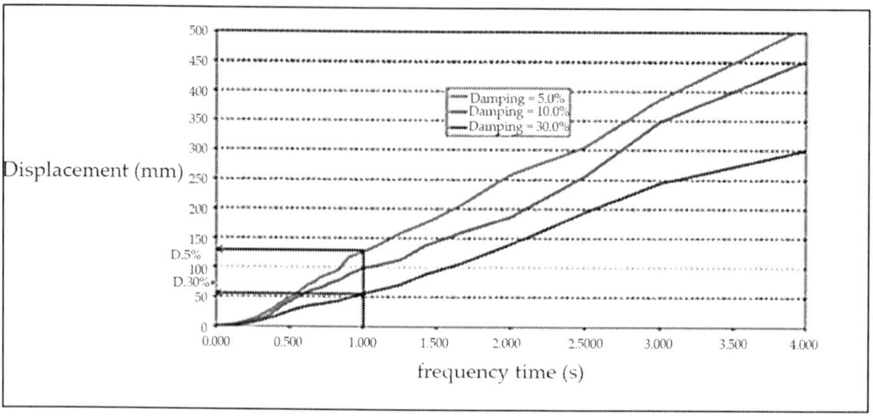

Figure 7.18: Displacement Vs Frequency

Types of Dampers

Dampers are classified based on their performance of friction, metal (flowing), viscous, viscoelastic; shape memory alloys (SMA) and mass dampers. Among the advantages of using dampers we can infer to high energy absorbance, easy to install and replace them as well as coordination to other structure members.(journal,2006).

Friction Dampers

In this kind of damper, seismic vitality is spent in beating erosion in the contact surfaces. Among different components of these dampers can be named maintaining a strategic distance from weariness in served loads (due to the non-dynamic dampers under load) and their execution autonomous to stacking speed and surrounding temperature. These dampers are introduced in parallel to propping (journal,2006).

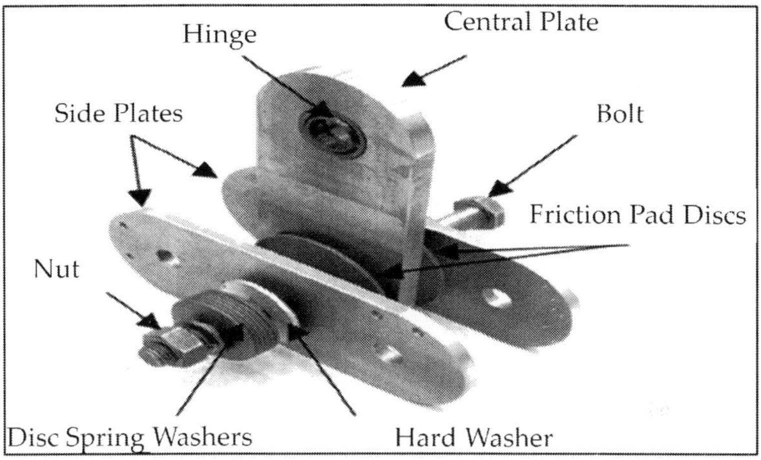

Figure 7.19: Friction Dampers

PVD Damper

It is another kind of contact damper and because of simplicity to establishment, is a standout amongst the most broadly utilized damper in structures (Warn, 2004). PVD damper can be utilized to make important damping for adaptable structures, for example, twisting steel outline or to give compelling damping to relative firmness of structures (Naeim, 1995). PVD damper is intended to establishment where removal can create essential damping, for example, establishment of metal skeleton prop or solid minute casing. The main building which was outlined by Iranian fashioner and by utilizing PVD damper of Robinson Company and it was a working with private proprietor in and with 164/5 squares meters territory and it will be implicit 6 stories. This building is situated in Rey. The floor is around 112 square meters. What's more, its fundamental skeleton is a sort of steel and screws and nuts sort. In this building, the retrofitting new innovation is utilized for tremor called seismic dampers framework. The quantities of PVD dampers are 12 damper of 100 KN. Another tall structure was composed in northern Tehran with 19 dampers of 350, 600, 850 KN PVD of Robinson organization by another creator which are in requesting stage(Huang et al., 2004).

Figure 7.20: PVD Damper

In hardware and ward dampers on the lead, for example, lead elastic orientation and lead dampers which are utilized as the best and best defensive gear for various structures against seismic tremor. Furthermore, they are developed in New Zealand nation by Robinson Company at DSIR material science and building research center. What's more, they have been utilized as the best dampers widely in most recent 30 years. PVD damper are composed based of lead plastic disfigurement, for example, lead elastic bearing. For elastic lead bearing, the made damping power by lead is not as much as versatile compel identified with elastic, while for PVD, damping power of lead is significantly more prominent than elastic flexible constrain. Presently, PVD is utilized as a viable damper on many scaffolds, towers, structures. The benefits of PVD damper incorporate (Hwang et al., 1994):

1. PVD damper acts viably on low removals. For instance, one 1MN PVD damper can acts successfully for 0.5 mm to 5 mm dislodging.
2. PVD damper requires no support and does not have any oil or winder segments.
3. PVD damper conduct resembles the conduct of a metal damper.

There are damper gear in which lead is utilized and they are introduced in thousand diverse structure the world over. In any case, ensuring high-raise structures with various volition with long stretches, for example, tall building is inconceivable because of regular frameworks or it is not financial. Robinson organization concocted PVD damper as an answer.

Also, it can decrease short and since quite a while ago settled structure. This damper can be accustomed to damping little vacillations and twist in tall structures. Likewise, we can decrease substitution brought about by quake. PVD is a damper can make a lot of hysteretic damping in little substitutions little to +/, 50 micrometer. By disseminating PVD damper in the structure, we can expand the size structure damping. This damper is appropriate to secure structure instability with a wide number of various periods. Lastly, it makes high damping for structure. Viable plan extend for a PVD test with a limit of 200 KN in view of got results incorporate that it is in the scope of +-50 micrometer to +-10mm. more noteworthy removal require a bigger sort yet its affectability will diminish.

Pall Friction Damper

Another kind of erosion damper is Pall grinding damper. This damper incorporates a supporting and some steel plate with contact screws. What's more, they ought to be introduced highly involved with supporting. Steel sheets are associated with each other by high quality jolts and they have a sneak past a specific compel, to each other.

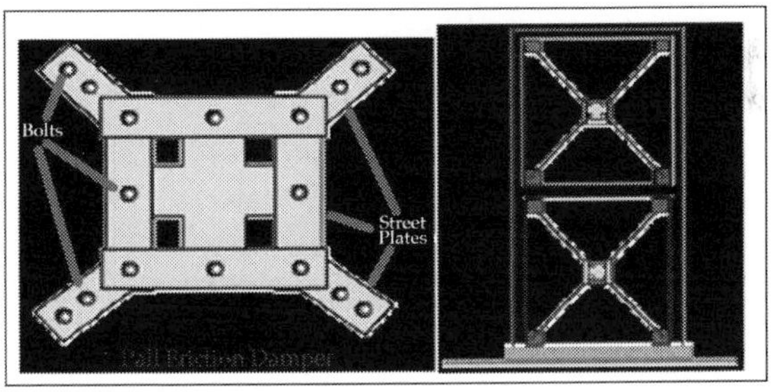

Figure 7.21: Pall grinding damper

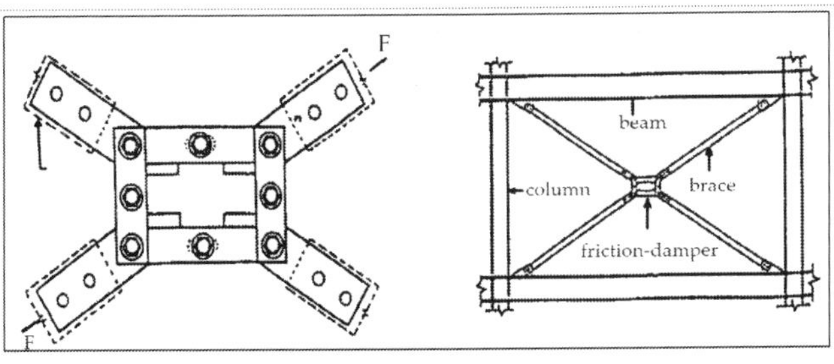

Figure 7.22: Pall Friction Damper

Metallic Dampers

In this damper, exchanged vitality to the structure is spent to accommodation and non-direct conduct in utilized component in damper. In these dampers, metal inelastic distortion is utilized, for example, for formability metals, for example, steel and lead for vitality dissemination. In every traditional structure, vitality scattering depends on misshapening of steel individuals after the accommodation.

Figure 7.23: Metallic Dampers

In supports, utilizing accommodation metallic dampers is more typical. These dampers are regularly made by some parallel steel plates. What's more, in mix with a propping framework, they embrace the part of assimilation and vitality dispersal. This piece of supporting can goes about as a breaker in structure. Furthermore, by concentrating on nonlinear conduct anticipate non-direct conduct and harm in other major and minor structure parts. X-formed metal dampers have a huge execution. Enormous accommodation on steel volume, giving Hysteretic damping and remarkable vitality dissemination are extraordinary components of this sort of damper. These dampers have a high sidelong solidness, notwithstanding giving damping. In this way, they were entitled as Added Damping And Stiffness (ADAS).

Led Injection Damper (LED)

This damper is made of a two-chamber cylinder, piston and lead inside piston. And by piston moving during earthquake, lead moves from larger chamber to smaller chamber. And with plastic deformation, the kinetic energy is wasted as heat. In figure 1-1-2-3, the longitudinal section of lead damper injection is shown (Saiidi et al.,1999).

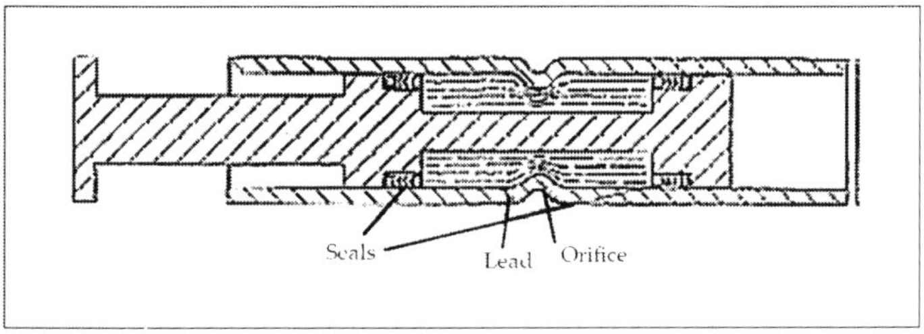

Figure 7.24: Led Injection Damper (LED)

Shape Memory Alloy

Shape Memory alloy (SMA) are created from metals which have the following properties:

Their flexibility is very similar to the flexibility of the rubber piece.

After applying many deformation, they can back to their original state, by heating.

The alloy of nickel and titanium has good resistance to corrosion, in addition to have these properties.

Viscous Dampers

In this damper, by utilizing gooey liquid inside a barrel, vitality is disseminated. Because of simplicity of establishment, flexibility and coordination with different individuals additionally assorted qualities in their sizes, gooey dampers have numerous applications in outlining and retrofitting.

Gooey dampers (VD) have been utilized as a part of the most recent 30 years in major common structures to relieve the impacts of quakes. Their utilization in elevated structures worked in seismic territories is a test for the creators, since they ought to diminish the vibrations actuated by both solid winds and quakes, and the ideal conduct in these two circumstances is not for the most part the same. Therefore, the outline prerequisite for VD to be utilized as a part of tall structures is frequently that they ought to have two distinct practices in the diverse scope of speeds relating to wind and quake. As of late VD with said conduct have been connected in three elevated structures in Asia, the Taipei 101 in Taipei, Taiwan, and the twin St. Francis Towers in Manila, Philippines.

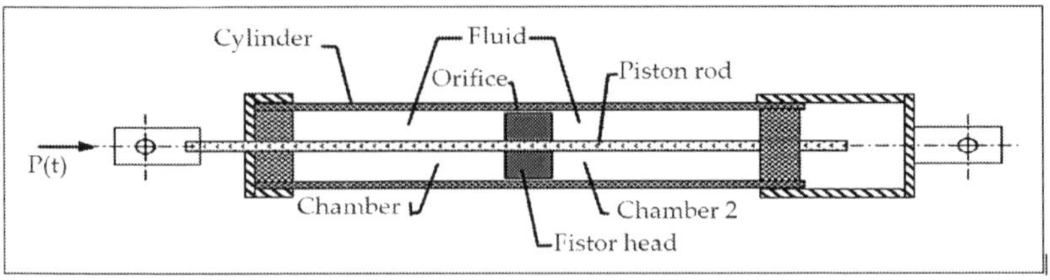

Figure 7.25: Viscous Dampers

This type of dampers are connected to the structure in three ways:

Damper installation in the floor or foundation (in the method of seismic isolation)

Connecting dampers in stern pericardial braces

Damper installation in diagonal braces.

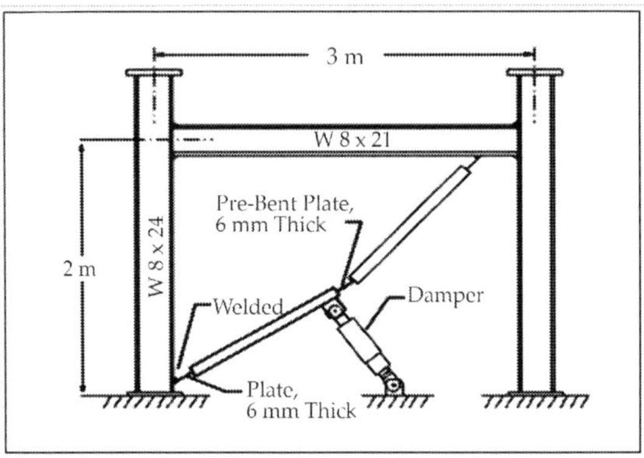

Figure 7.26: Viscous Dampers

Mass Tuned Dampers

Mass is placed on a fulcrum which acts as a roller. And it allows to mass with move as a transfer-lateral movement to the floor. Springs and dampers are placed between mass and anchor members to the floor and frame and they are placed relative in "opposite phase" and sometimes are adjacent vertical. And these anchor members transmits structural lateral force. Bidirectional transfer dampers are made as a spring-damper in two vertical directions. And they provide controlling the structure movement in two vertical structures. Seismic isolator is used to isolate the structures of strong ground motions during an earthquake. Unlike building which its isolation is often done on the foundation, this separation is applied between topside of structure and below side of that in bridges. This is due to a high inertia force of topside (including deck weight) also it will be easy to use. The main objective in seismic isolation is to reduce base frequency of structure vibration and reaching it to a lower value than frequencies which have main earthquake energy. In other word, seismic isolation increases structure vibration period and bridge and distances it from periods containing earthquake main energies. So, the input energy to base caused by earthquake is reduced with seismic isolation. Another advantage of seismic isolation is to provide a tool to waste energy. So, inputted energy to structure get wasted in small points and by a controlled manner. Thus, destruction and damaging in particular and few concentrated points will be exist and it will be possible to replace these parts after the earthquake. In general, seismic isolation design lead to reduce the structures responses reduction which are in earthquake conditions.

Figure 7.27: Mass Tuned Dampers

- Increasing main period
- Increasing relative damping (energy dissipation)

Isolation objectives in a bridge is quite different with a building. In a building, due to reducing the energy forces applied to topside and to reduce stress, isolators are installed in structural elements. But in a separate bridge, seismic isolator is installed to keep the elements under isolators (backpacks and base) to reduce the transmitted energy force and change topside places (deck) to below side structures (backpacks and base).

Lead Rubber Damper LRM and Rubber Damper HDRB

We use isolator in bridge design in order to achieve maximum energy absorption (relative to the period of the isolated structures). So, we should use isolators which have a high damping. And from this perspective, LRB seismic isolator has an intrinsic property 30% (independent of vibration frequency, temperature and environmental conditions).

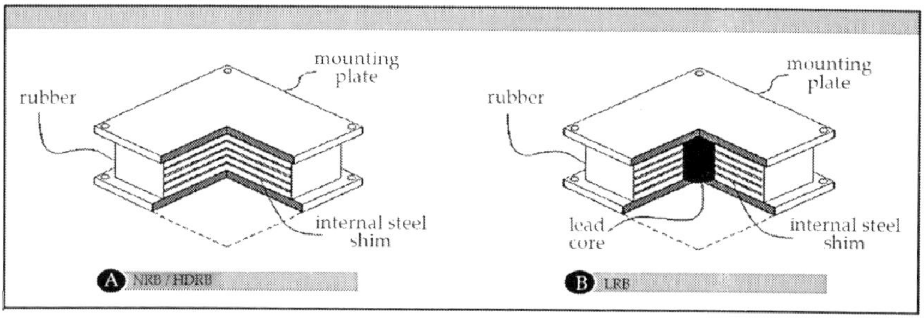

Figure 7.28: Lead Rubber Damper LRM and Rubber Damper HDRB

For bridges designing because of the movement limitation, the effect of increasing alternative period is low in the most projects, but the damping effect in bridges is higher than construction projects.

Comparing LRB with HDRB:

LRB will have damping to 30% but HDRB will have damping between 10 % to 15 %.

Start LRB movement, due to this fact that first lead should flows from stiffness to flowing phase, it has a greater time delay than HDRB.

HDRB price is less than LRB price.

HDRB has a lower resistant movement against weaker earthquakes and wind than LRB due to the initial lead stiffness.

HDRB have more sensitivity to environmental vibrations than LRB.

LRB performance relative to earthquake is better than HDRB.

NEW BREED OF ENERGY DISSIPATION DEVICES

The imaginative techniques for control of seismic vibrations, for example, frictional and different sorts of damping gadgets are critical vital piece of seismic segregation frameworks as they serious as a boundary against the infiltration of seismic vitality into the structure. In this idea, the dampers smother the reaction of the segregated building with respect to its base.

The novel grinding damper gadget comprises of three steel plates turning against each other in inverse bearings. The steel plates are isolated by two shims of grinding cushion material delivering grating with steel plates.

At the point when an outer compel energizes a casing structure the brace begins to dislodge evenly because of this drive. The damper will take after the movement and the focal plate in view of the tractable strengths in the propping components. At the point when the connected powers are turned around, the plates will pivot in inverse way. The damper scatters vitality by methods for contact between the sliding surfaces(Bertero, 2002).

FRICTION-VISCO ELASTIC DAMPER DEVICE (F-VEDD)

The latest **Friction-Visco Elastic Damper Device (F-VEDD)** combines the advantages of pure frictional and viscoelastic mechanisms of energy dissipation. This new product consists of **friction pads** and **viscoelastic polymer pads** separated by steel plates. A pre-stressed bolt in combination with disk springs and hardened washers is used for maintaining the required clamping force on the interfaces as in original FDD concept.

SEISMO-RESISTANT BUILDING DESIGN AND ARCHITECTURE

The carelessness of the association between the safe structure and other non-basic components in the building adjusts the resistance-solidness relationship of auxiliary components. In such a circumstance, it is conceivable that the structure does not demonstrate all the safe limit at the same time as required amid the seismic activity. This venturing of the building's seismo-safe limit brings about its halfway harm or an aggregate crumple. This is one purpose behind the unadulterated basic examination getting to be noticeably inadequate to attest the seismic resistance of a building. Subsequently, an extra approach winds up plainly vital, which may see the seismo-safe framework overall, where all the basic, non-basic and space-shaping components accommodating a building are thought to connect with each other and consider in charge of the seismo-safe limit. As indicated by this approach, such obligation is shared by both the auxiliary examination and plan. Indeed, it is not sought after here to acquire from the figuring engineer a quantitative assessment of positive and negative impacts of all the associating components of a building, but instead to match them with the basic plan. For example, if a sismo-safe structure or inflexible edge is comprised of a given number of sections, they ought to all demonstrate the seismo-safe limit at the same time amid the seismic stage. Something else, if attributable to any reason, not respected in the Structural Design and Analysis, simply some portion of them act toward the start of the shake, this will be inadequate and in the end, will separate. The staying ones will bomb thus, achieving the crumple of the building. This venturing does not allude to such plastic distortions which might be balanced by methods for a sufficient redistribution of requesting and stresses. It alludes to extreme venturing where such plastic distortions are amazingly overpowered.

General Objectives

To prevent seismo-resistance stepping.

To optimize the seismo-resistance.

Seismo-Resistant Design Theory

The development of our SRA was done on four study lines.

To schematize and improve the Seismo-resistant Structural Design under the constraints of the Architectural Design.

To define the general principles of the Seismo-resistant Building Architecture.

To schematize and improve the interrelations and compatibilization between the Structural Design and the Seismo-resistant Architectural Design.

To develop a SRA based upon the congruent interrelations between every and each interacting structural subsystem (or component).

Basic Principle

The hypothesis, technique and research which will help build up that approach ought to unavoidably meet the accompanying necessities of this fundamental rule: "The seismo-safe auxiliary components ought to create the firmness, quality malleability and synchronization as foreseen in the basic plan and investigation when subjected to seismic activity".

That is, they ought to have the capacity to demonstrate their seismo-safe limit all the while. Something else, the safe limit may venture up, and this in the long run may bring about the structure's disappointment. From the basic examination perspective, and for the instances of firm chunks or floors, this is accomplished by circulating the seismic shear relatively to each safe component's solidness and torque corresponding to firmness, and separation to the focal point of torsion. In general, and as an essential element, the basic and engineering plans ought to accomplish the vital similarity among all the basic components of the working, with a specific end goal to fulfill this fundamental rule.

Basic Criteria

The accompanying four fundamental criteria for Seismo-safe Structural Design are legitimate in the SRA too, to keep the seismo-resistance venturing, as well as to streamline the outline.

1. The seismic coefficient for the different stories of a building increments as indicated by the building tallness. Thus, amid the compositional outline, it is imperative to place documents, swimming pools, or rooms containing substantial gear in lower levels, keeping in mind the end goal to limit seismic impacts.
2. The safe components might be set with a specific level of autonomy from the vertical load. This element strikingly encourages both the auxiliary and the engineering

plans. Truth be told, it enables us to put the primary safe components in the most appropriate mold to limit the torsional impacts and to go to building venture prerequisites.

3. Seismic powers are relative to the building weight. With seismic strengths corresponding to the building weight, it is prudent to decrease the last mentioned. By and by, this objective can be accomplished in the accompanying things:

- Partition dividers Floor Pieces Floors Wall putting as well as covering Resistant Structure Building Typology

This standard, straightforward as it might appear, is by the by disregarded in many building ventures, even these days.

Effectiveness and Optimization of the Seismo-resistant Response in Buildings

Basically, this target might be accomplished either diminishing the seismic strengths or expanding the effectiveness of the seismic limit of structures. Decrease of estimations of seismic strengths might be accomplished in different ways, i.e.:

1. By utilizing lightweight materials or staying away from pointless fillings and finishings.
2. By migrating heavier weights, that is, attempting to put those rooms that will bear heavier weights (e.g. chronicles, swimming-pools, meeting rooms, and so on.) in lower levels. Seismic twisting minutes and shearing following up on the structure are in this way lessened and subsequently, the span of the safe components. These are critical actualities for the Architectural Design.
3. By staying away from the pseudo-reverberation. This way to keep the key time of the working from harmonizing with the fundamental one of the establishment soil.
4. The advancement of the seismo-safe limit of the building must be finished by utilizing spatial shapes that may prompt a working with an unmistakable and basic

structure, having its torsion push focus incidental with its mass focus.

This joined motivation behind viability and improvement is an unmistakable test for the Architecture, since it includes to study particular philosophies which empower the building configuration to make critical commitments for the best answer for the seismic issue.

Reasons Which Cause the Stepping of the Seismo-resistant Capacity of a Building

Explanations behind which the most extreme seismic tremor safe limit of a building is not equivalent to the whole of the limits of every component.

1. Inherent causes to the auxiliary plan: The procedure for coordinating both the resistance and the firmness of basic components experiences trials with troublesome arrangements in the tremor basic outline hone, which presents an instability element that may offer ascent to a decline in seismic tremor safe limit qualities, and a later fall of the structure.
2. Earthquake torsion: The way that the seismic shear in a section brought about by a torsion minute is corresponding to the separation to the focal point of torsion or solidness (CT) brings about, for segments which are dimensionally equivalent, however put at various separations from CT, their seismic anxieties are distinctive.

- This might be the central reason for the diminishing in safe limit, particularly if the torsional impacts have not been accommodated.
- This might be the reason for the venturing up of the seismic resistance, particularly if the torsional impact has not been enough accommodated. There are some different situations where the improvement of a minute is because of an unpredicted strain of floors.
- Finally, notwithstanding when torsion is considered into the basic investigation, it is as yet hard to keep the solidness resistance proportion for sections.

3. Flexible story: An adaptable floor significantly diminishes the seismic shear on the staying stiffer floors, be that as it may, in the meantime, the shear increments on the adaptable floor itself. Here, once more, we have a venturing up of seismic resistance. Considering this reality amid the auxiliary examination calls for pliability and resistance attributes which are extremely hard to accomplish in the practice. It is ideal to allot the vital solidness resistance proportion to such a story as opposed to utilizing it to enhance the seismo-safe execution of the entire building.
4. Short pillars: With unbendingly confined structures, when one of the bars on a specific floor or level is of a quite lesser length than the staying ones, the supposed short bar case happens. Here the issue emerges in light of the fact that rakish solidness is contrarily corresponding to its length. Like in the past cases, the trouble here is to accomplish the required resistance-solidness relationship. The centralization of bowing minutes may break the shaft, encouraging the abatement of the seismo-safe limit of the structure.
5. Non-basic components: It is referred to that non-basic components, for example, dividers, parcel dividers, establishments, and so forth meddle with the normal conduct of the safe structure. This obstruction might be both of a positive or a negative sense. This has likewise been a regular explanation behind diminishing the building all out seismic limit.
6. Constructive imperfections: A limited valuable deformity on the safe structure, other than diminishing the seismo-safe limit, may start venturing. Truth be told, it might offer ascent to a sudden torsion minute.

Like productive deformities, this might be the reason for a reduction in the seismo-resistance limit and its surprising venturing.

Standards of Compatibility of Architectural and Structural Designs

The referential variables which are to be morphologically compatible are considered at the very beginning.

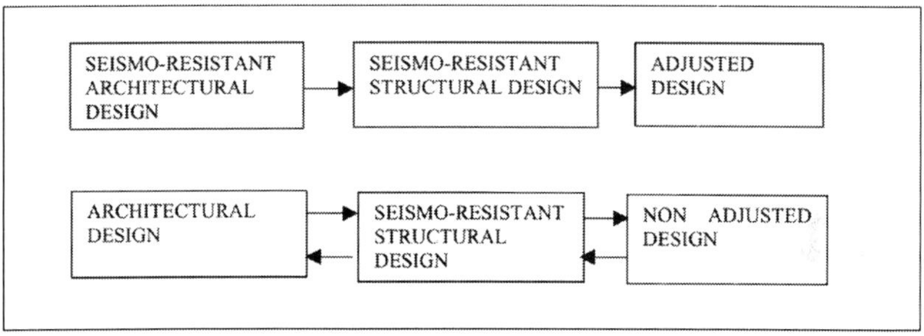

Figure 7.2: Objectives (1) Adjusted Design (2) Non-Adjusted Design

- Flexible floor
- Building Collision
- Seismic Torsion
- Pseudo-resonance
- Sudden stiffness changes in plan and elevation
- Stiffness-Flexibility
- Concentrated weight
- Short columns
- Tall buildings
- L- U- and T-shaped buildings

Now, the morphological answers for each variable. The following table is a first attempt which would admit an immediate application of the suggested solution.

Flexible Floors

This circumstance emerges when, at a specific floor, the solidness of a tall building is extensively lessened in connection to the bordering floors. This circumstance causes a solid convergence of seismic powers on the site, offering ascend to an unsafe venturing instrument of the building resistance. The morphological answer is to maintain a strategic distance from this element in the building plan. At whatever point a story with extensive detachments between segments is required, it ought to be the last one or it ought to be put outside the tower site, ideally outlined as a solitary level(Murty, Goswami, Vijayanarayanan, & Mehta, 2012).

Building Collision

This wonder happens when there are no joints between adjoining structures and the crash is created when the motions are not synchronized. This is totally irregular which must be completely maintained a strategic distance from. The morphological answer is building detachment, as present tenets indicate. It is prescribed to mull over into the outline the different elements of the totally isolated bodies for the instance of a similar working, keeping in mind the end goal to forestall building crash, to give a uniform structure and furthermore to evade sudden firmness changes in plan and height.

Seismic Torsion

This impact is created at whatever point the Stiffness Center (SC) and the Torsion Center (TC) don't harmonize, in this way bringing about extra limitations particularly in those components which are far expelled from the SC, which may prompt the venturing of the seismo-safe limit of the building.

In spite of the fact that this issue is considered in the auxiliary examination, it is totally undesirable since it produces substantial extra and misbalanced seismic strengths in the arrangement of segments, offering ascend to the venturing of the seismo-safe limit of the building.

The morphological arrangement is met by planning structures with a symmetrical arrangement and height. Also, the auxiliary and non-basic communicating components symmetry is required, and also the practical symmetry of the design site.

Pseudo-Resonance

This marvel emerges at whatever point the time of the building matches the overwhelming time of the establishment soil. This condition astoundingly builds the seismic impacts. Then again, if the principal vibration time of the building relies on upon its measurements and structure firmness, then, the morphological arrangement is to control these parameters.

Stiffness-Flexibility

At whatever point an unbending or an adaptable building is required, i.e. one that can be stressed to a specific low or high degree separately, the normal practice is to use for the primary case inflexible structures, for example, segment dividers comprised of strengthened cement and additionally high thickness and high resistance stone work dividers, 0.20m. thick, and for the second case, the determination is for materials which are satisfactory for adaptable structures. Both cases will impact the spatial morphology of such structures.

Concentrated Weight

In the greater part of the current seismo-safe measures, the seismic coefficient increments relatively to the floor level concerning the ground level. Thusly, in Architectural Design, this guideline must be borne at the top of the priority list, not exclusively to abstain from utilizing overwhelming materials, subfloors, segment dividers, coverings., at more elevated amounts, additionally to put the locales intended for documents, swimming-pools, or substantial gear at lower levels. In this manner, two objects are accomplished: right off the bat, a decrease of the seismic strengths, since the seismic coefficient increments at more elevated amounts and also, a consistent diminishment of the seismic shear and minutes.

The accompanying case obviously delineates the significance of this later idea.

It displays a six-level development for looking at the seismic impact brought on by a specific P weight, which is right off the bat set at the fifth level and after that at the principal level of a similar development. (Figure 2)

The outcomes are decisive. On account of P set at the fifth level, the toppling minute moves toward becoming 25 times more noteworthy than that for P set at the principal level. Furthermore, the seismic shear influences to levels 1 up to 5, though, in the second case, just the primary level is influenced yet to a lesser degree (5 times less)(Bertero, 2002).

Short Columns

Another perspective identified with the resistance-firmness issue is the supposed "Short Column". For this situation, the seismic shear increments conversely relative to the 3D shape of its stature for segments of equivalent cross-sectional region. Furthermore, this circumstance intensifies for short segments since cement is unsatisfactory for opposing solid distracting stresses, hence strikingly diminishing its flexibility.

These examples are begun by a specific component of brick work which decreases the segments stature and therefore, their solidness turns out to be extraordinarily expanded. This makes the seismic shear focus on the section, which legitimately can't help it. The crack of the safe components could make whatever remains of the components yield too, which could thus realize the aggregate fall. This circumstance can be effortlessly kept away from by suitably planning the shape and area of spaces and openings. Then again, when this issue comes about because of contrasts in rise between medium-tallness mezzanines, its end is basically inconceivable. Thusly, these height contrasts must be expelled from the seismo-safe engineering plan(Murty et al., 2012).

REFERENCES

1. Adams, R. (2002). International seismology. INTERNATIONAL GEOPHYSICS SERIES, 81(A), 29-38.
2. Bertero, V. (2002). Innovative approaches to earthquake engineering.
3. Bolt, B. A. (2004). Engineering seismology Earthquake Engineering: From Engineering Seismology to Performance-Based Engineering: CRC Press.
4. Bozorgnia, Y., & Bertero, V. V. (2004). Earthquake engineering: from engineering seismology to performance-based engineering: CRC press.
5. Chandra, S., & Krishna, G. (1983). Effect of tannery waste disposal on the quality of the river Ganges in Kanpur. Poll. Res, 2, 63-64.
6. Charleson, A., & Taylor, M. (2004). Earthquake architecture explorations.
7. Coe, J. A., Michael, J. A., Crovelli, R. A., Savage, W. Z., Laprade, W. T., & Nashem, W. D. (2004). Probabilistic assessment of precipitation-triggered landslides using historical records of landslide occurrence, Seattle, Washington. Environmental & Engineering Geoscience, 10(2), 103-122.

8. Dolan, J. F., & Rockwell, T. K. (2001). Paleoseismologic Evidence for a Very Large (Mw> 7), Post-AD 1660 Surface Rupture on the Eastern San Cayetano Fault, Ventura County, California: Was This the Elusive Source of the Damaging 21 December 1812 Earthquake? Bulletin of the Seismological Society of America, 91(6), 1417-1432.
9. Elnashai, A. S., & Di Sarno, L. (2008). Fundamentals of earthquake engineering: Wiley New York.
10. Gioncu, V., & Mazzolani, F. (2002). Ductility of seismic resistant engineering structures: Spon Press.
11. Gioncu, V., & Mazzolani, F. (2010). Earthquake engineering for structural design: CRC Press.
12. Goretti, A., & Di Pasquale, G. (2004). Building inspection and damage data for the 2002 Molise, Italy, earthquake. Earthquake Spectra, 20(S1), S167-S190.
13. Henger, M., Berckhemer, H., & Seidl, D. (2002). The history of the development of the German Regional Seismic Network. Ten years of German Regional Seismic Network (GRSN). Report, 25, 1-8.
14. Holzer, T. L., Scawthorn, C., & Rojahn, C. (2005). Coordinating NEHRP Post-Earthquake Investigations—Exercising the Plan. Earthquake Spectra, 21(4), 1043-1062.
15. Hu, Y.-X., Liu, S.-C., & Dong, W. (1996). Earthquake engineering: CRC Press.
16. Huang, Y., Bertero, V. V., & Wada, A. (2004). Innovative strategies in earthquake engineering Earthquake engineering: From engineering seismology to performance-based engineering: CRC Press.
17. Kanamori, H. (2005). Real-time seismology and earthquake damage mitigation. Annu. Rev. Earth Planet. Sci., 33, 195-214.
18. Lee, W. H., Jennings, P., Kisslinger, C., & Kanamori, H. (2002). International handbook of earthquake & engineering seismology (Vol. 81): Academic Press.
19. Lomnitz, C., & Elizarrarás, S. R. (2001). El Salvador 2001: earthquake disaster and disaster preparedness in a tropical volcanic environment. Seismological Research Letters, 72(3), 346-351.
20. Minor, R., & Grant, W. C. (1996). Earthquake-induced subsidence and burial of late Holocene archaeological sites, northern Oregon

coast. American Antiquity, 772-781.

21. Mucciarelli, M., Masi, A., Gallipoli, M. R., Harabaglia, P., Vona, M., Ponzo, F., & Dolce, M. (2004). Analysis of RC building dynamic response and soil-building resonance based on data recorded during a damaging earthquake (Molise, Italy, 2002). Bulletin of the Seismological Society of America, 94(5), 1943-1953.
22. Murty, C., Goswami, R., Vijayanarayanan, A., & Mehta, V. V. (2012). Some concepts in earthquake behaviour of buildings. Gujarat State Disaster Management Authority, Government of Gujarat.
23. Oskin, M., Sieh, K., Rockwell, T., Miller, G., Guptill, P., Curtis, M., . . . Elliot, P. (2000). Active parasitic folds on the Elysian Park anticline: Implications for seismic hazard in central Los Angeles, California. Geological Society of America Bulletin, 112(5), 693-707.
24. Raup, D. (1997). Perils of a restless planet: Scientific perspectives on natural disasters: by E. Zebrowski, Jr Cambridge University Press, 1997.£ 16.95 hbk (xi+ 306 pages) ISBN 0 521 57374 2: Elsevier Current Trends.
25. Spector, T. (1997). Ethical dilemmas and seismic design. Earthquake Spectra, 13(3), 489-504.
26. Sucuoğlu, H., & Akkar, S. (2014). Basic Earthquake Engineering: Springer.
27. Wiegel, R. L. (1970). Earthquake engineering Earthquake engineering: Prentice-Hall.
28. Yeats, R. S., & Gath, E. M. (2004). The Role of Geology inSeismic Hazard Mitigation Earthquake Engineering: From Engineering Seismology to Performance-Based Engineering: CRC Press.

INDEX

M

N

O

P

Q

R

S

T

U